QCE Units 3 & 4

BIOLOGY

WENDY COOK

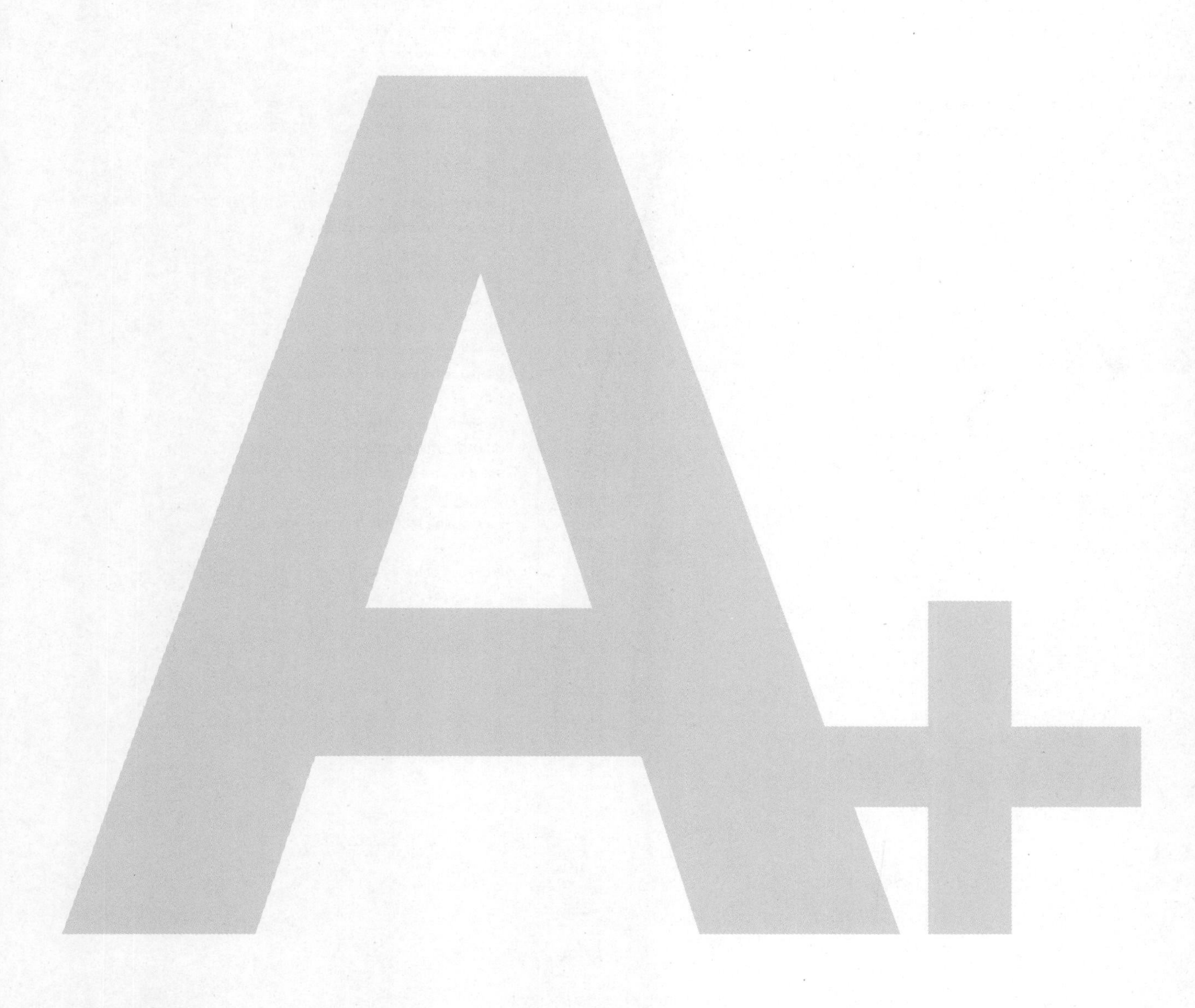

- topic summary notes
- exam practice questions
- detailed, annotated solutions
- study and exam preparation advice

STUDY NOTES

A+ Biology QCE Units 3 & 4 Study Notes Workbook
1st Edition
Wendy Cook
ISBN 9780170459136

Publisher: Sarah Craig
Project editor: Felicity Clissold
Cover design: Nikita Bansal
Text design: Alba Design: Rina Gargano
Project designer: Nikita Bansal
Permissions researcher: Wendy Duncan
Production controller: Karen Young
Typeset by: SPi Global

Any URLs contained in this publication were checked for currency during the production process. Note, however, that the publisher cannot vouch for the ongoing currency of URLs.

Acknowledgements
© State of Queensland (QCAA) 2019 https://www.qcaa.qld.edu.au/senior/senior-subjects/sciences/biology/syllabus CC BY 4.0 https://creativecommons.org/licenses/by/4.0/

For product information and technology assistance,
in Australia call **1300 790 853**;
in New Zealand call **0800 449 725**

For permission to use material from this text or product, please email **aust.permissions@cengage.com**

ISBN 978 0 17 045913 6

Cengage Learning Australia
Level 7, 80 Dorcas Street
South Melbourne, Victoria Australia 3205

Cengage Learning New Zealand
Unit 4B Rosedale Office Park
331 Rosedale Road, Albany, North Shore 0632, NZ

For learning solutions, visit **cengage.com.au**

Printed in China by 1010 Printing International Limited.
1 2 3 4 5 6 7 25 24 23 22 21

CONTENTS

UNIT 3 BIODIVERSITY AND THE INTERCONNECTEDNESS OF LIFE

9780170459136

UNIT 4 HEREDITY AND CONTINUITY OF LIFE

HOW TO USE THIS BOOK

A+ Biology QCE Units 3 & 4 Study Notes is designed to be used year-round to prepare you for your QCE Biology exam. *A+ Biology QCE Units 3 & 4 Study Notes* includes comprehensive topic summaries of all key knowledge in the QCAA Biology General Senior syllabus that you will be assessed on during your exam. Chapters 1 & 2 and 4 & 5 address each of the topics in the syllabus. Chapter 3 specifically addresses the data test, which is a separate assessment task worth 10% of your final mark. This section gives you a brief overview of each chapter and the features included in this resource.

Topic summaries

The topic summaries at the beginning of each chapter give you a high-level summary of the essential subject matter for your exam.

Concept maps

The concept maps at the beginning of each topic provide a visual summary of the hierarchy and relationships between the subject matter in each topic.

Key knowledge summaries

Key knowledge summaries in each chapter address all key knowledge of the syllabus. Summaries are broken down into sequentially numbered chunks for ease of navigation. Step-by-step worked examples and 'hints' unpack the content and help to prevent mistakes.

Glossary

All your key terms for each topic are bolded throughout the key knowledge summaries and included in a complete glossary at the end of the chapter. Digital flashcards are accessible via QR code and provide a handy revision tool.

Revision summary

The revision summaries are a place for you to make notes against each of the syllabus dot points to ensure you have thoroughly reviewed and understood the content.

Exam practice

Each topic concludes with multiple-choice and short-response questions to test your recall of the key concepts and to practise answering the types of questions you will face in your exam. Selected questions from the QCAA 2020 Biology exam papers have been included for authentic exam practice. An icon alerts you to these questions. ©QCAA QCAA 2020 P1 MC Q3

Complete solutions to practice questions are available at the back of the book to provide immediate feedback and help self-correct errors. They have been written to reflect a high-scoring response. This section includes explanations for multiple-choice answers, and explanations to short-response items demonstrating high-scoring responses, with mark breakdowns, and signposts to potential mistakes.

A+ DIGITAL FLASHCARDS

Revise key terms and concepts online with the A+ Flashcards. Each topic glossary in this book has a corresponding deck of digital flashcards you can use to test your understanding and recall. Just scan the QR code or type the URL into your browser to access them.

https://get.ga/aplus-qce-bio-u34

Note: You will need to create a free NelsonNet account.

ABOUT THE AUTHOR

Wendy Cook

Wendy Cook has been teaching in Science for 24 years. She is currently the Head of Science at an independent school on the Sunshine Coast, where she has a focus on developing skills and scientific thinking in students.

Wendy has taught in both the UK and Australia, including A Level, OP and ATAR Biology. She has written curriculum programs and internal assessment items and has had many years of experience as a QCAA Panellist. She is now a QCAA Assessor for Biology.

PREPARING FOR THE EXAM

Exam preparation is a year-long process. It is important to keep on top of the theory and consolidate your knowledge regularly, rather than leaving revision to the last minute. Revise often, choosing one or two dot points to focus on. You should aim to have the theory learnt and your notes complete so that by the time you reach study leave, the revision you do is structured, efficient and meaningful.

Study tips

To stay motivated to study, try to make the experience as comfortable as possible. Have a dedicated study space that is well lit and quiet. Create and stick to a study timetable, take regular breaks, reward yourself with social outings or treats and use your strengths to your advantage. For example, if you are a visual learner, turn your Biology notes into cartoons, diagrams or flow charts. If you are better with words or lists, create flashcards or film yourself explaining tricky concepts and watch it back.

Revision techniques

Here are some useful revision methods to help make information **STIC**.

Spaced repetition	This technique uses the Leitner method, which helps to move information from your short-term memory to your long-term memory by spacing out the time between your revision and recall sessions; for example, using flashcards you have created. As you slowly extend the time for retrieving information, your brain processes and stores the information for longer periods.
Testing	Testing is necessary for learning and is a proven method for exam success. The 'hypercorrection effect' shows that when you are confident of an answer that is actually incorrect, and the answer is corrected after feedback, you are more likely to remember the correct answer, thereby improving your future performance. Further, if you test yourself before you learn all the content, your brain becomes primed to retain the correct answer when you get it.
Interleaving	Interleaving is a revision technique that sounds counterintuitive but is very effective for retaining information. Most students tend to revise a single topic in a session, and then move onto another topic next session. With interleaving, you choose three topics (1, 2, 3) and spend 20–30 minutes on each topic. You then study 1,2,3 or 2,1,3 or 3,1,2 'interleaving' the topics, and repeating the study pattern over a long period of time. This strategy is most helpful if the topics are from the same subject and are closely related.
Chunking	An important strategy is breaking down large topics into smaller, more manageable 'chunks' or categories. Essentially, you can think of this as a branching diagram or mind map where the key theory or idea has many branches coming off it that get smaller and smaller. By breaking down the topics into manageable chunks, you will be able to revise the topic systematically.

These strategies take cognitive effort, but that is what makes them much more effective than re-reading notes or trying to cram information into your short-term memory the night before the exam!

Time management

It is important to manage your time carefully throughout the year. Make sure you are getting enough sleep, that you are getting the right nutrition, and that you are exercising and socialising to maintain a healthy balance so that you don't burn out.

To help you stay on target, plan out a study timetable. One way to do this is to:

1. Assess your current study and social activities. How much time are you dedicating to each?
2. List all your commitments and deadlines, including sport, work, assignments, family and so on.
3. Prioritise the list and re-assess your time to ensure you can meet all your commitments.
4. Decide on a format, whether it be weekly or monthly, and schedule in a study routine.
5. Keep your timetable somewhere you can see it.
6. Be consistent.

The exam

The end of year examination accounts for 50% of your total mark. It assesses your achievement in the following objectives of Biology Units 3 & 4:

1 describe and explain biodiversity, ecosystem dynamics, DNA, genes and the continuity of life, and the continuity of life on Earth
2 apply understanding of biodiversity, ecosystem dynamics, DNA, genes and the continuity of life, and the continuity of life on Earth
3 analyse evidence about biodiversity, ecosystem dynamics, DNA, genes and the continuity of life, and the continuity of life on Earth to identify trends, patterns, relationships, limitations or uncertainty
4 interpret evidence about biodiversity, ecosystem dynamics, DNA, genes and the continuity of life, and the continuity of life on Earth to draw conclusions based on analysis.

The examination includes two papers. Each paper includes several possible question types:
- multiple-choice questions
- short response items requiring single-word, sentence or paragraph responses
- calculating using algorithms
- interpreting graphs, tables or diagrams
- responding to unseen data and/or stimulus
- extended response (300–350 words or equivalent).

You will have 90 minutes plus 10 minutes perusal time for each paper.

The day of the exam

The night before your exam, try to get a good rest and avoid cramming, as this will only increase stress levels. On the day of the exam, arrive at the venue of your exam early and bring everything you will need with you. Rushing to the exam will increase your stress levels and will reduce your ability to do well. Further, if you are late, you will have less time to complete the exam, which means that you may not be able to answer all the questions or you may rush to finish and make careless mistakes. Don't worry too much about 'exam jitters'. A certain amount of stress is required to help you concentrate and achieve an optimum level of performance. However, if you are still feeling very nervous, breathe deeply and slowly. Breathe in for a count of six seconds, and out for six seconds until you begin to feel calm.

Perusal time

Use your time wisely! *Do not* use the perusal time to try and figure out the answers to any of the questions until you have read the whole paper. The exam will not ask you a question testing the same knowledge twice, so look for hints in the stem of the question and avoid repeating yourself. Plan your approach so that when you begin writing you know which section, and ideally which question, you are going to start with.

Strategies for effective responses

Pay particular attention to the cognitive verb used in the question. For example, a question with the cognitive verb 'explain' requires a different response from a question with the cognitive verb 'describe'. Familiarise yourself with the definitions of the commonly used cognitive verbs on the next page (listed in order of complexity). Understanding the definitions of these cognitive verbs will help ensure you are not just providing general information or restating the question without answering it.

Describe	give an account (written or spoken) of a situation, event, pattern or process, or of the characteristics or features of something
Explain	make an idea or situation plain or clear by describing it in more detail or revealing relevant facts; give an account; provide additional information
Apply	use knowledge and understanding in response to a given situation or circumstance; carry out or use a procedure in a given or particular situation
Analyse	dissect to ascertain and examine constituent parts and/or their relationships; break down or examine in order to identify the essential elements, features, components or structure; determine the logic and reasonableness of information; examine or consider something to explain and interpret it, for the purpose of finding meaning or relationships and identifying patterns, similarities and differences
Interpret	use knowledge and understanding to recognise trends and draw conclusions from given information; make clear or explicit; elucidate or understand in a particular way; bring out the meaning of, e.g. a dramatic or musical work, by performance or execution; bring out the meaning of an artwork by artistic representation or performance; give one's own interpretation of; identify or draw meaning from, or give meaning to, information presented in various forms, such as words, symbols, pictures or graphs

Multiple-choice questions

Read the question carefully and underline any important information to help you break the question down and avoid misreading it. Read all the possible solutions and eliminate any clearly incorrect answers. Fill in the multiple-choice answer sheet carefully and clearly. Check your answer and move on. Do not leave any answers blank.

Short-response questions

It is important that you plan your response before writing. To do this, **BUG** the question:

- **B**ox the cognitive verb (describe, explain, apply etc.).
- **U**nderline any key terms or important information, and take note of the mark allocation.
- **G**o back and read the question again.

Many questions require you to apply your knowledge to unfamiliar situations so it is okay if you have never heard of the context before, but you should know which part of the syllabus you are being tested on and what the question is asking you to do. If a stimulus is included, use information from that as part of the response to show how you are linking the (unfamiliar) context to your knowledge.

Plan your response in a logical sequence. If the question says 'describe and explain' then structure your answer in that order. You can use dot points to do this, but ensure you write in full sentences. Rote-learnt answers are unlikely to receive full marks, so you must relate the concepts of the syllabus back to the question and ensure that you answer the question that is being asked. Planning your response to include the relevant information and the key terminology will help you to not write too much, contradict yourself, or 'waffle' on and waste time. If you have time at the end of the paper, go back and re-read your answers.

Good luck. You've got this!

UNIT 3 BIODIVERSITY AND THE INTERCONNECTEDNESS OF LIFE

3

Chapter 1
Topic 1 Describing biodiversity

Topic summary

Biodiversity encompasses the interactions and processes that occur in different environments with all species of organisms, including variants and genes adapted to different climates, that result in different ecosystems.

Biodiversity refers to the measurement of Earth's biosphere on an increasing scale from individuals through populations, species, communities and ecosystems. It considers genetic variation within species and between species and diversity in habitats.

There are two common ways of measuring biodiversity:

- measuring the difference or similarity of species between studied areas (e.g. recording species richness)
- measuring the abundance of species in an area.

There need to be sustainable interactions between species for the biosphere and its associated ecosystems to survive. Over time, the relationships that exist have changed and evolved and they can be classified in many different ways.

There are several techniques for collecting data to measure diversity and interactions within and between species:

- Quadrats are used to measure abundance, density or percentage cover.
- Transects are used to measure distribution.
- Belt transects are a combination of line transects and quadrats and are used to measure both abundance and distribution.

Sampling techniques can be random, systematic or stratified.

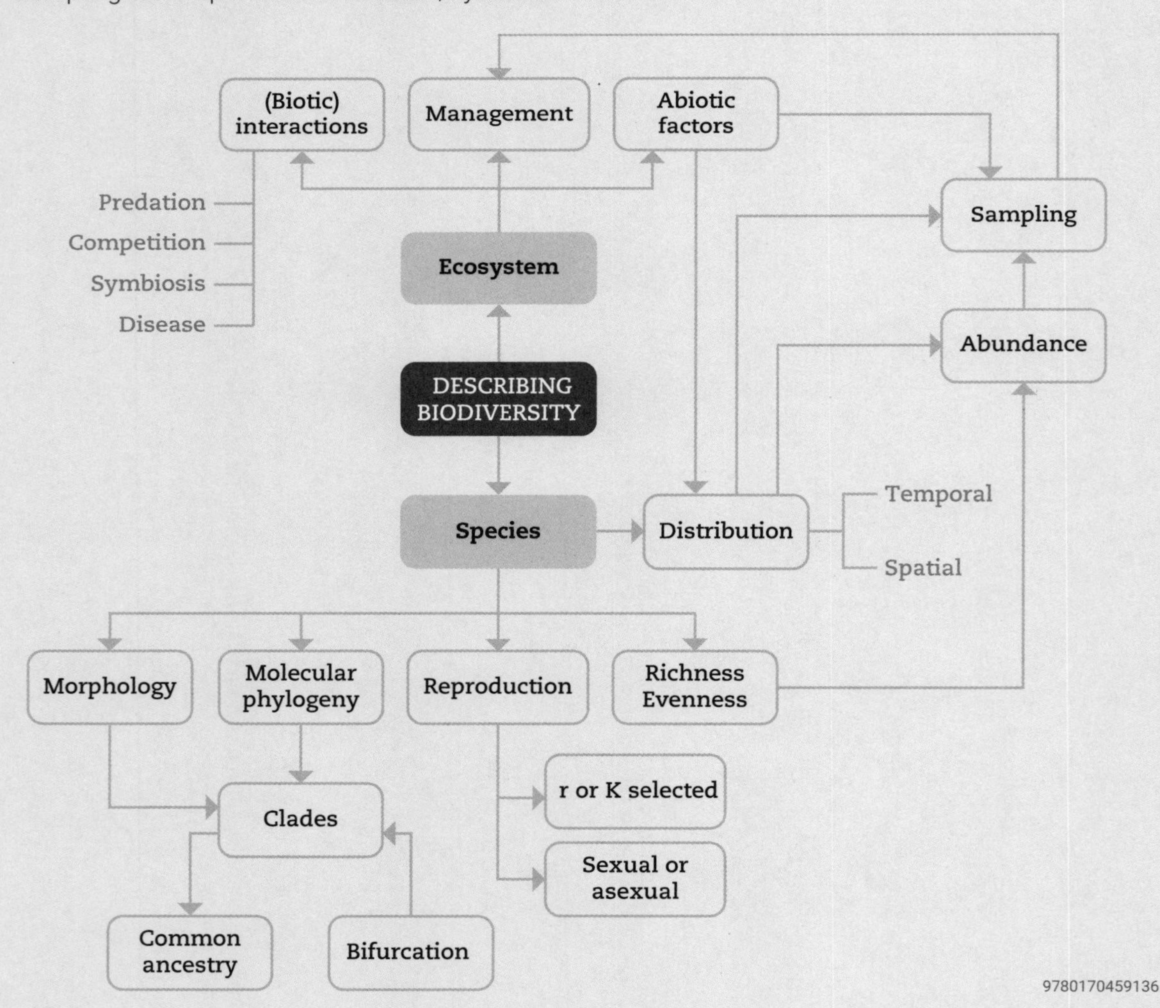

1.1 Biodiversity

1.1.1 Diversity of species

Diversity can be measured in several ways. Each method gives different information about the populations, species and/or communities being researched. Researchers recording this information can identify similarities and differences between research areas. Measures of diversity include species richness, species evenness and species diversity. Each measure provides different information, so it is important to have a clear understanding of each method and choose the best one to answer the question being researched.

Species richness refers to the number of types of species in a designated area. It is easy to count different species. However, species richness has limitations because it does not consider abundance.

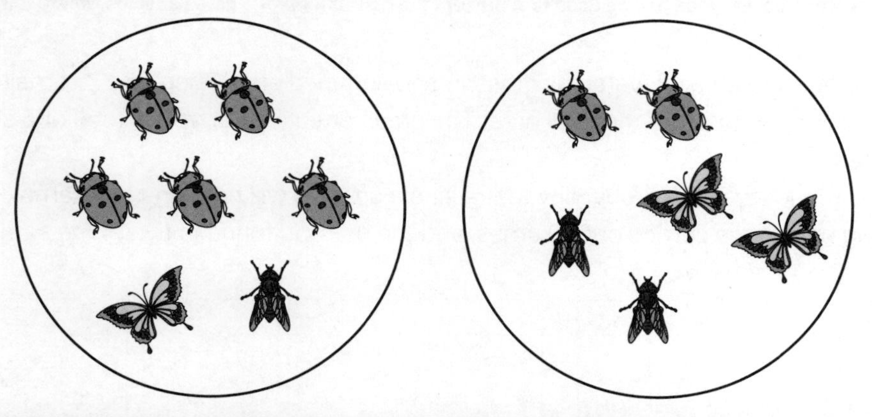

FIGURE 1.1 The two samples have the same species richness but different species evenness.

Another limitation when using species richness is the effect of sampling size – in small samples, some species may be missed (Figure 1.2).

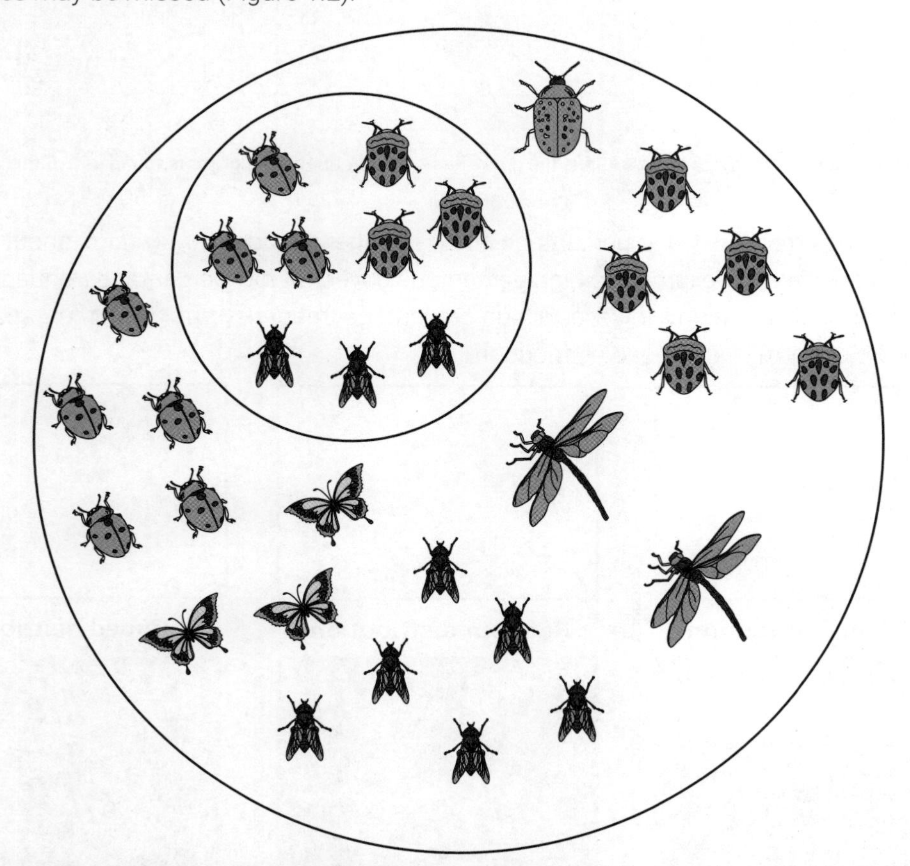

FIGURE 1.2 Some species can be missed with smaller sample sizes.

9780170459136

Hint
This links to sampling techniques.

Species evenness refers to the relative abundance of each of the species in the designated area.

Other measures of abundance are **percentage cover** and **percentage frequency**. Percentage cover is used when it is difficult to identify individual organisms of a species, such as grasses. It is based on the amount of space one species takes up.

FIGURE 1.3 Percentage cover: grids can be used to represent the distribution of a grass species, where it is hard to count individuals.

Figure 1.3 shows the areas covered by grass in one sampled area. There are 25 small squares in the grid, each representing 4% of the total area. The grass effectively covers a total of 5 squares or 20% of the area.

Percentage frequency is the probability that a single species will be found in a sample area. If the above sampling of grass is carried out 10 times, and the grass is found in five of the sample areas, then it has a percentage frequency of 50%.

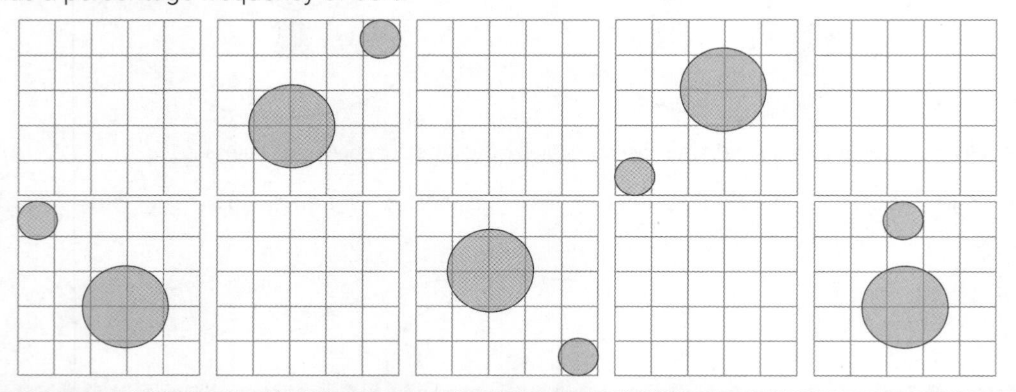

FIGURE 1.4 Percentage frequency: in this example, the grass has a 50% frequency as grass is found in 5 of the 10 sample areas.

Using percentage frequency and percentage cover together provides information about the distribution of a species. For example, a high percentage cover but low percentage frequency suggests a clumped distribution. However, a high percentage frequency but low percentage cover suggests a more uniform, widespread distribution.

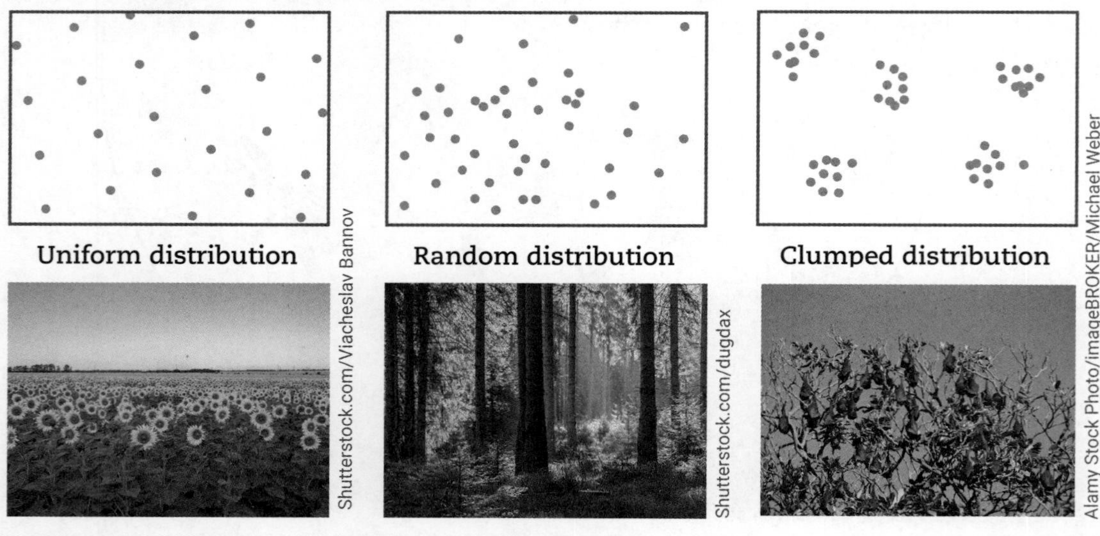

FIGURE 1.5 Organisms can have uniform, random or clumped distribution.

Species diversity considers both types (richness) and relative abundance (evenness) of species. The syllabus uses Simpson's diversity index (SDI) to calculate diversity.

$$\text{SDI} = 1 - \left(\frac{\sum n(n-1)}{N(N-1)}\right)$$

where *n* is the number of organisms of one species and *N* is the total number of organisms of all species. Note: *D* may also be used to represent Simpson's diversity index, but it is best to use SDI as the abbreviation.)

$n(n - 1)$ relates to the relative abundance of species. $N(N - 1)$ relates to the number of different types of species. SDI looks at how likely it is that individuals in the sample are *different*. Therefore, the value is subtracted from 1. SDI ranges from 0 (low diversity) to 1 (high diversity).

The sample from the small circle in Figure 1.2 (page 3) gives an SDI of 0.60. Data from the larger circle in Figure 1.2 is summarised in Table 1.1.

TABLE 1.1 Summarising the data from the large circle in Figure 1.2

Species	Count (*n*)	*n*(*n* – 1)
Bug 1	5	5 × 4 = 20
Bug 2	3	3 × 2 = 6
Bug 3	6	6 × 5 = 30
Bug 4	2	2 × 1 = 2
Bug 5	5	5 × 4 = 20
Bug 6	1	1 × 0 = 0
	N = 22	$\sum n(n-1) = 78$
$\text{SDI} = 1 - \left(\frac{\sum n(n-1)}{N(N-1)}\right)$	$\text{SDI} = 1 - \frac{78}{22(22-1)}$ $= 1 - \frac{78}{462}$ $= 1 - 0.169$ $= 0.831 = 0.83$	

Hint
You will need to know what each 'n' represents. Remember that the small *n* is for each individual species, whereas the big *N* is for all the species together.

Hint
It is important to substitute all values into the equation, even the zero values. Round for decimal places at the end of your calculation too.

Note that you should not use only SDI to make judgements about a community. A high species diversity may result from the presence of undesirable species. A community may have low species diversity but include some rare, or highly desired, species. It is important to consider the SDI in the context of the species identified.

1.1.2 Species interactions

Species do not live in isolation. They live in an ecological community and interactions between individuals of each species form the basis of nutrient cycling and food webs. Understanding these interactions also helps you determine an organism's abundance and distribution on a temporal and spatial scale. The most consistently occurring interactions across a range of communities are symbiosis, competition, predation and disease.

Intraspecific interactions occur between individuals of the same species, whereas **interspecific interactions** occur between individuals of two or more species. These interactions can be used to classify organisms (refer to these definitions when you get to Unit 3 Topic 1 Classification processes when classifying according to species interactions).

Competition occurs when individuals require a common resource that is in limited supply. The outcome usually has negative effects on the weaker competitors. **Competitive exclusion** occurs when a weaker competitor is eliminated from an area by a stronger competitor. Coexistence of species occurs when intraspecific competition is stronger than interspecific competition because each species inhibits their own population growth before they inhibit the population growth of the competitor.

Predation occurs when one individual kills and eats another individual of a different species. This can include herbivores 'preying' on plants, although this tends to be called grazing.

When two or more species live in close association with each other for a long time, the interaction is termed **symbiosis**. The three major types of symbiosis are mutualism, commensalism and parasitism.

In **mutualism**, both individuals benefit from the association; mutualism is represented as + +. **Commensalism** is represented as + 0, because one individual benefits and the other neither benefits nor is harmed. **Parasitism** occurs when the parasite benefits while causing harm to the host, and is represented as + –. Generally, the parasite does not kill the host and may not complete its entire life cycle within the host.

Like parasitism, **disease** occurs when a pathogen lives in or on a host organism, causing harm to that organism. Pathogens that cause disease can be transmitted between individuals of a species or a community. Pathogens reproduce inside their host in a very short time compared to the lifespan of the host, producing an infection that results in immunity or death of the host. Pathogens include bacteria, fungi, many protozoa and viruses.

TABLE 1.2 Examples of interactions between organisms

Predator The bilby catches and eats insects.	Copyright Greg Harold/ AUSCAPE All rights reserved
Mutualism + + Cape sugarbirds feed on the nectar from the king protea flower then transfer pollen to other plants. The sugarbird gets food and the protea is pollinated.	Alamy Stock Photo/ David G Richardson
Commensalism + 0 Dung beetles process animal manure and return nutrients to the soil. The beetle gets food and the animals are unaffected.	Shutterstock.com/PRILL
Parasitism + – *Microporus xanthopus* fungus takes nutrients away from the *Acacia* tree. Eventually, the tree dies because of lack of nutrients.	iStock.com/teptong

››

›› **Interspecific competition** Lions and hyenas are different species competing for the same resource.	 Alamy Stock Photo/Nature Picture Library/Anup Shah
Intraspecific competition Members of the same species compete for the same resource.	 iStock.com/David Clarke

1.1.3 Abiotic factors

An **abiotic factor** is a physical or non-living component that shapes the environment of an ecosystem. Abiotic factors can affect the abundance and distribution of organisms. Climate is an abiotic factor that describes the atmospheric conditions in an area over a long period. Another abiotic factor is substrate – the surface (e.g. soil) on which an organism grows. The area investigated may be terrestrial or aquatic.

Abiotic factors can be measured; for example:

- temperature of air, water and soil is measured in °C
- humidity is measured as a percentage (%)
- salinity is often measured in % or ppm of water or soil
- pH of water or soil is measured on the pH scale
- light intensity is measured in lumens.
- precipitation is measured by volume (mm).

Table 1.3 lists some abiotic factors that have been measured for the Tanami Desert in northern Australia (Figure 1.6).

TABLE 1.3 Summary of abiotic factors for the Tanami Desert

Abiotic factor	Summer	Winter
Rainfall (mm)	298 (monsoonal)	–
Humidity (%)	25	40
Average maximum temperature (°C)	36–38	25
Average minimum temperature (°C)	20–22	10
Wind speed (km h^{-1})	12	12

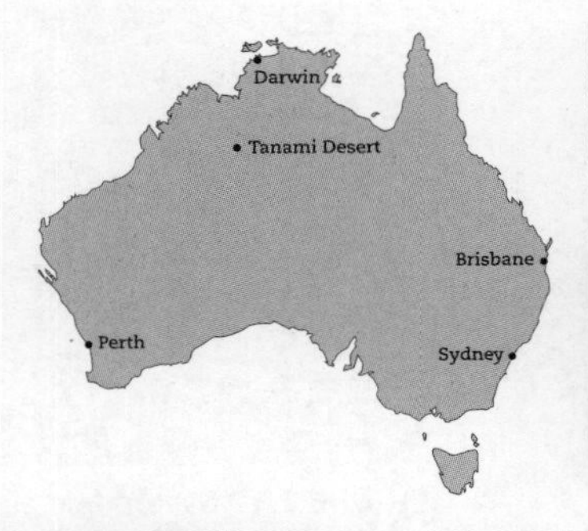

iStock.com/Patrick Honan

FIGURE 1.6 The Tanami Desert

1.1.4 Sampling

It is not practical to count every individual or association in a community. So, when ecologists investigate diversity and interactions, they collect data on a portion of the entire area or population. They sample a representative area and collect information about diversity, species interactions and abiotic factors. This can be used to compare ecosystems across spatial (different areas) and temporal (different times or days or years) scales.

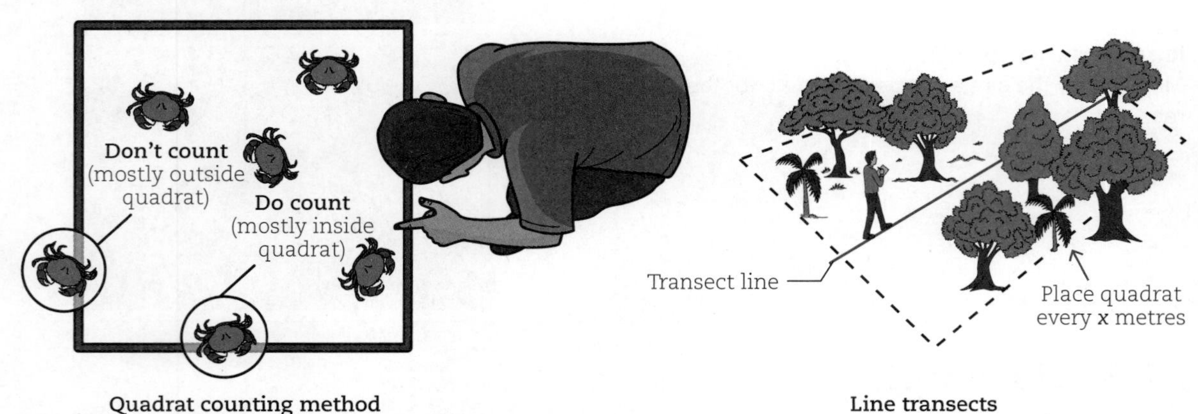

FIGURE 1.7 Two common ways of sampling ecosystems are quadrats and line transects.

Random sampling is used when an area being investigated is fairly uniform (with evenly spaced organisms) or very large and/or time is limited. Each part of the sampled area has an equal chance of being chosen, reducing bias. Usually this involves the use of quadrats (Figure 1.8). **Quadrats** give information about species type and abundance. The organisms inside the quadrat are counted individually or the percentage cover is measured. The size of the quadrat is based on the size of the organism and the sample area. Quadrats can be difficult to use with large organisms such as trees.

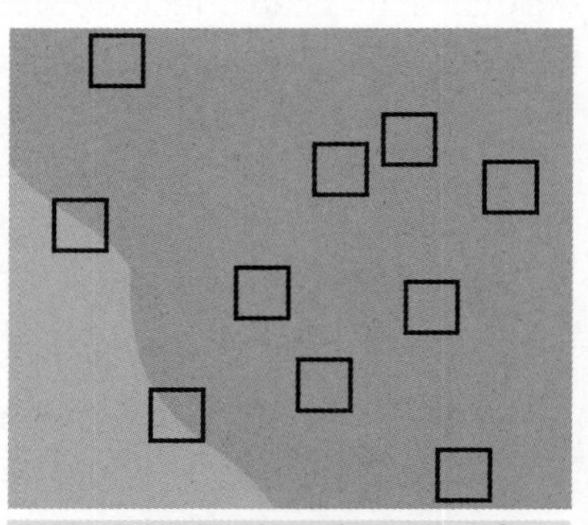

FIGURE 1.8 Random placement of quadrats

Systematic sampling is used when there is an environmental gradient in the sample area. Usually this involves the use of set intervals along a transect line.

Line transects give information about species types along a gradient (Figure 1.9). This allows for the changes taking place to be visualised. Line transects do not give information about species abundance. Individuals touching the line are recorded continuously or at set intervals, along with the factor associated with the gradient.

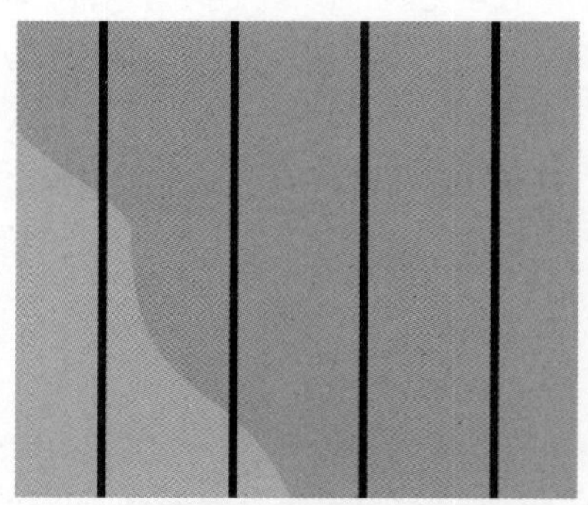

FIGURE 1.9 Systematic placement of line transects

A **belt transect** (essentially a series of quadrats) is used if information is required about both distribution and abundance of species (Figure 1.10).

Transects are run perpendicular to the sample area margin. Points along the line are placed either randomly or uniformly, depending on the question being asked. In this way, the distribution of a species across (or within) the environmental gradient can be determined.

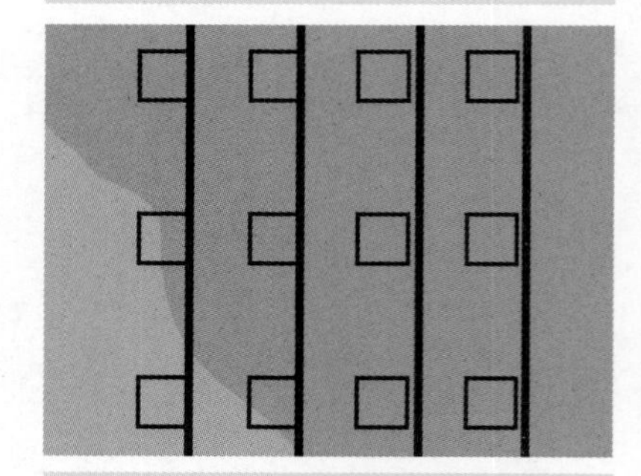

FIGURE 1.10 Systematic placement of belt transects

Stratified sampling is used when there are different strata (areas that are different from each other) in the sample area and a proportionate number of samples is taken from each area (Figure 1.11).

Figure 1.12 shows you how to decide which sampling method to use.

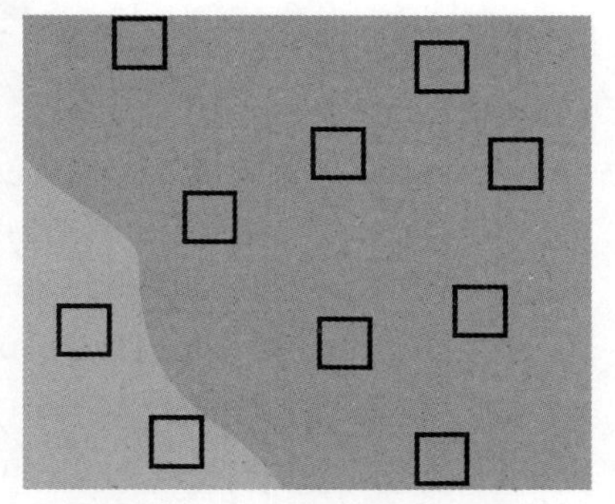

FIGURE 1.11 In this sample, 20% of the area is one type of habitat; therefore, two of the 10 quadrats are placed randomly in that stratum.

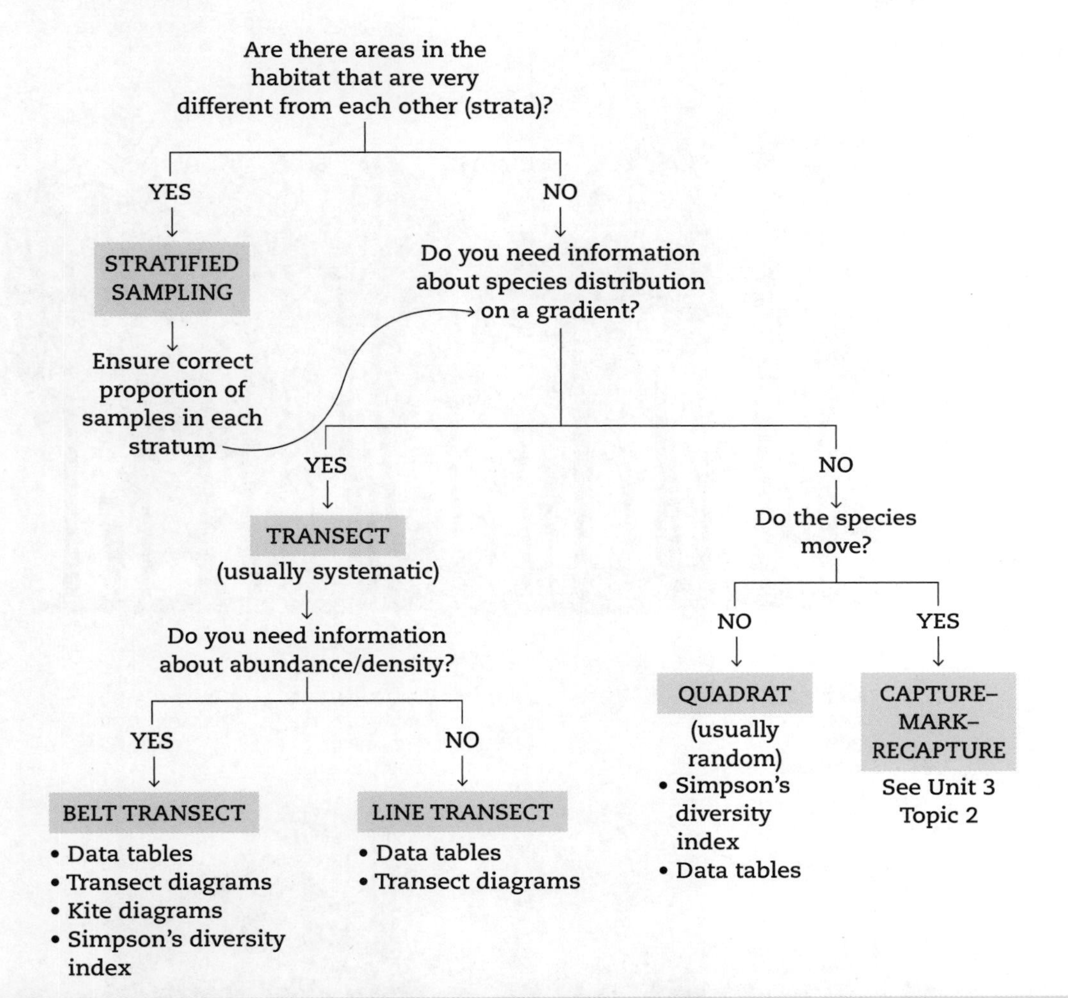

FIGURE 1.12 Steps to take when deciding how to sample a particular area. It is important to determine whether the data collected will allow you to answer the research question and whether the technique can be used in the sample environment.

1.1.5 Comparing ecosystems

Once you have collected data about ecosystems, it can be processed and organised to compare ecosystems. Table 1.4 (page 10) shows data for abiotic and **biotic factors** for dry and wet rainforests in Queensland. Table 1.5 (page 12) shows the number and type of species recorded during sampling. The table includes an SDI calculation for both dry and wet rainforest sites.

TABLE 1.4 Abiotic and biotic factors in dry and wet rainforest habitats in Queensland

Factor		Dry rainforest	Wet rainforest
Abiotic factor	Average maximum temperature (°C)	20.5	28.6
	Average minimum temperature (°C)	13	22.5
	Average humidity %	65	67
	Carbon storage potential (t)	170–800	300–1700
	Average annual rainfall (mm)	600–1200	1000–2000
Biotic factor	Canopy height (m)	4–30	≥30
	Canopy cover (%)	30–70	>70
	Leaf shedding	Yes, when dry	No

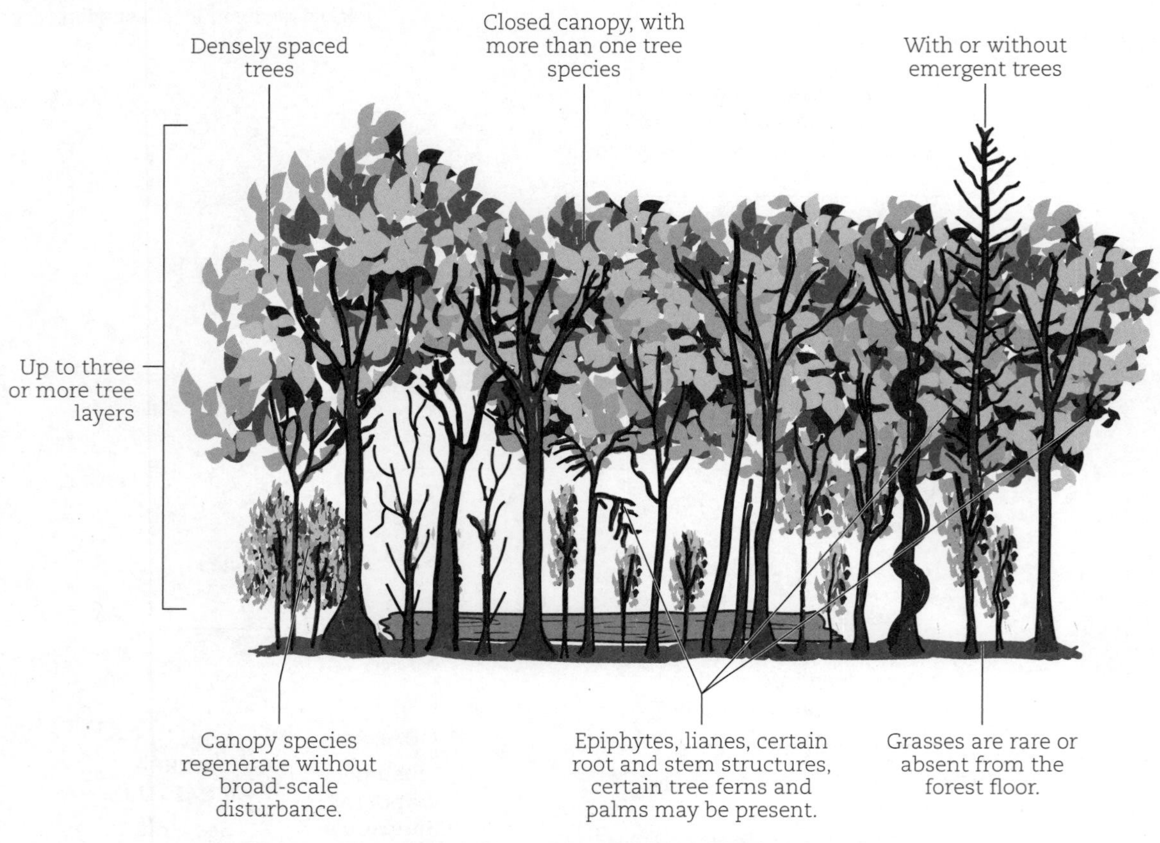

Shutterstock.com/AustralianCamera

FIGURE 1.13 The appearance and structure of a wet rainforest

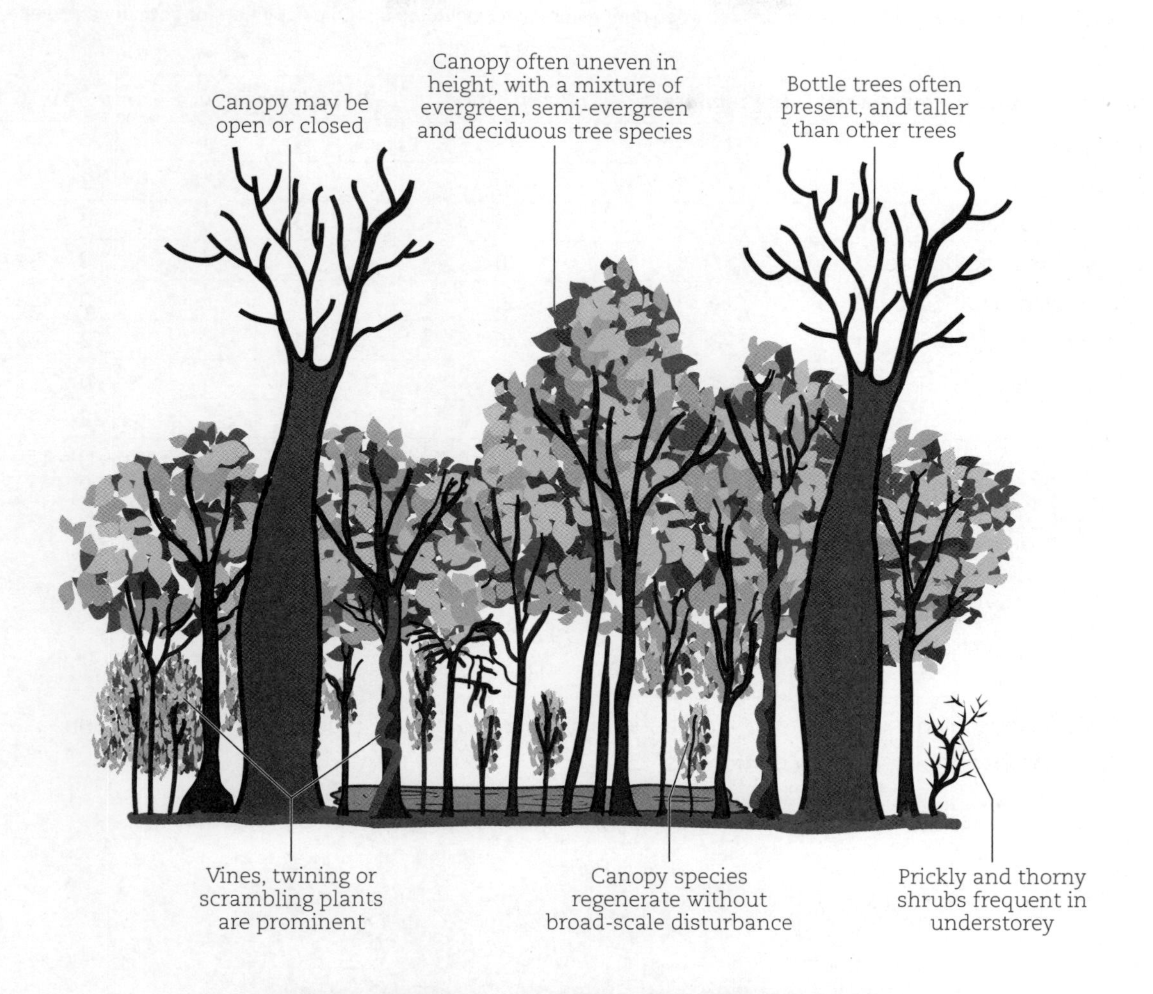

Alamy Stock Photo/Art Directors & TRIP

FIGURE 1.14 The appearance and structure of a dry rainforest

TABLE 1.5 The number and type of species recorded during sampling, including an SDI calculation for both dry and wet rainforests

Plant species in a sample area	Dry rainforest	$n(n-1)$	Wet rainforest	$n(n-1)$
Hoop pine	10	90	0	0
Lantana	6	30	4	12
Ribbonwood tree	0	0	10	90
Narrow-leaved bottle tree	5	20	0	0
Small-leaved scrub ironbark	6	30	0	0
White cedar	3	6	4	12
Hairy bird's eye	2	2	3	6
Ground orchid	0	0	6	30
	$N = 32$	$\Sigma n(n-1) = 178$	$N = 27$	$\Sigma n(n-1) = 150$
$\text{SDI} = 1 - \left(\frac{\Sigma n(n-1)}{N(N-1)}\right)$	$\text{SDI} = 1 - \frac{178}{32(32-1)}$ $= 1 - \frac{178}{992}$ $= 1 - 0.179$ $= 0.820 = 0.82$		$\text{SDI} = 1 - \frac{150}{27(27-1)}$ $= 1 - \frac{150}{702}$ $= 1 - 0.214$ $= 0.786 = 0.79$	

A good way to organise points for comparison is a Venn diagram. Figure 1.15 shows a Venn diagram comparing wet and dry rainforests.

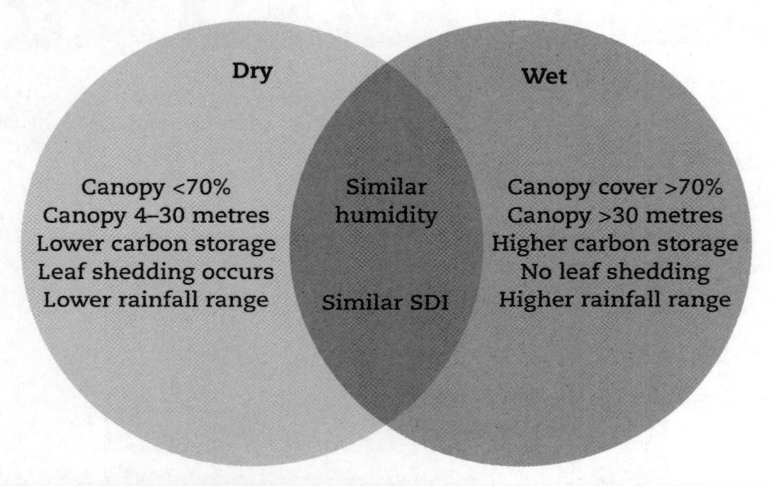

FIGURE 1.15 A Venn diagram identifying similarities and differences between dry and wet rainforests

Hint
Remember that 'compare' means you need to identify similarities, differences *and* the significance of these.

From Figure 1.15, we can make the following comparison. The lower rainfall averages in the dry rainforest (*difference*) probably contribute to the lower canopy height and cover because the types of trees that can grow do not form a complete canopy as do those in the higher rainfall area of wet rainforests (*difference effect*). Even though there are differences in the structure of the community, there is a similar species richness and evenness at each of the sample areas (*similarity*), resulting in a similar SDI value (*similarity effect*).

1.1.6 Factors affecting distribution and abundance of species

Hint
Look back at how to calculate SDI – show your working and substitute *all* the values.

Species are often capable of moving physically between ecosystems. Barriers to wider **distribution** are usually related to changes in temperature, water, precipitation or soil type. These factors, combined with species interactions (such as the presence or absence of prey), limit the dispersal of most species. For example, eastern grey kangaroos live in the area around Urangan, which is on the coast of Queensland near Hervey Bay and is where ferries for Fraser Island depart and arrive. Eastern grey kangaroos can travel at speeds of 24 km h^{-1} but they cannot travel the 27 km to Fraser Island because the water presents a barrier. Similarly, the arid conditions of deserts prevent any species that need high levels of water from moving to or across this habitat.

Observe the relationship between Australian climate zones and the distribution of taipans shown in Figures 1.16–1.18.

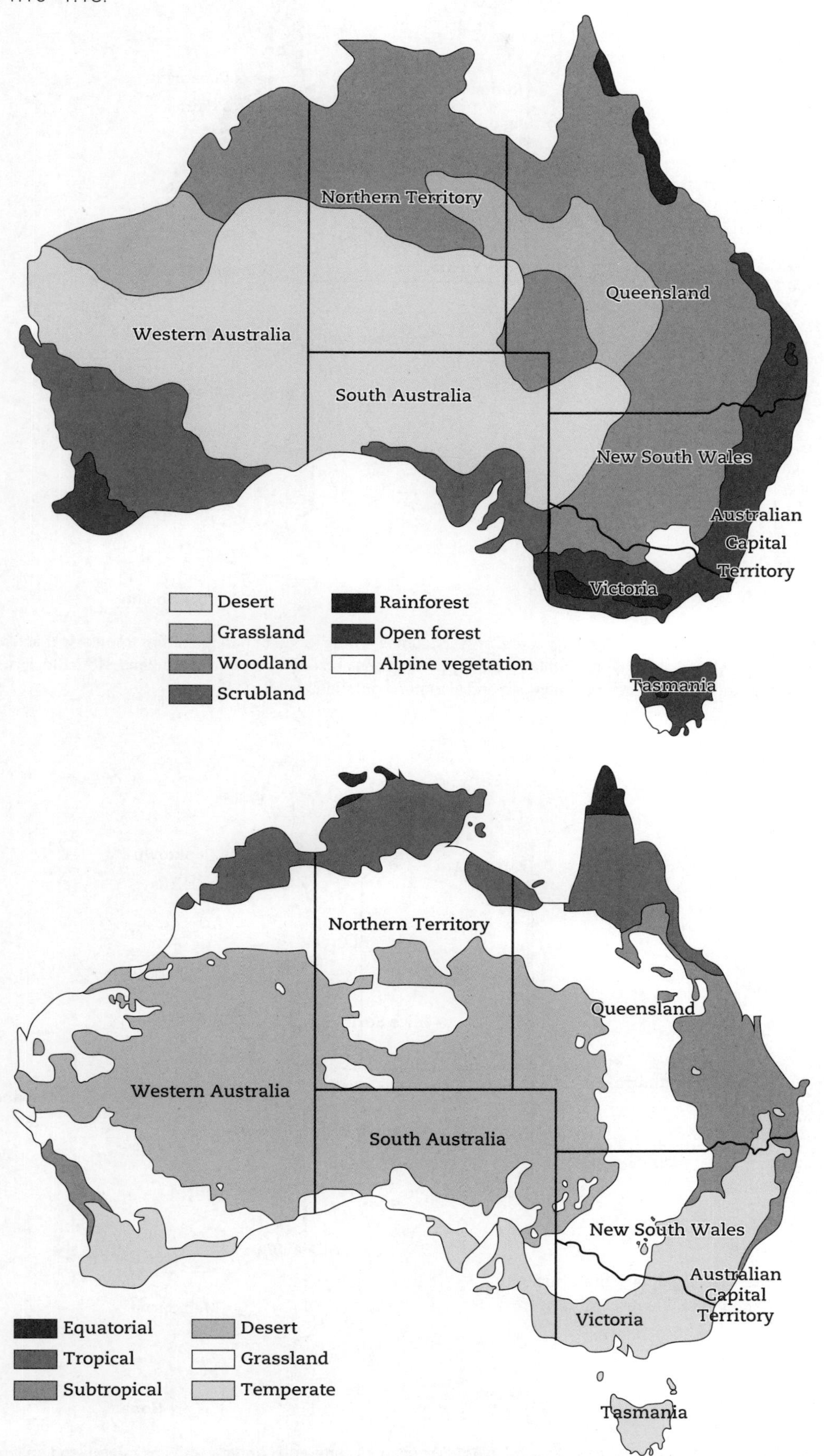

FIGURE 1.16 Australian vegetation and climate zones

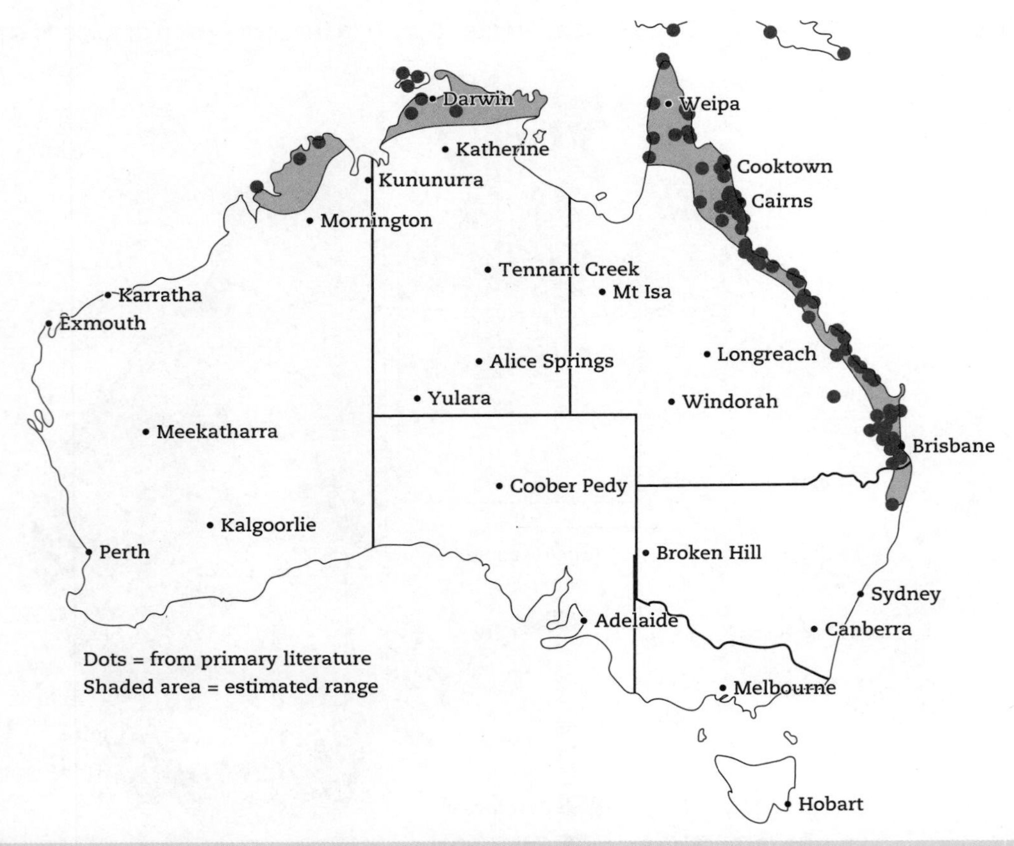

FIGURE 1.17 The coastal taipan (*Oxyuranus scutellatus*) occupies a wide range of habitats, from tropical wet sclerophyll forest to dry sclerophyll forest, coastal heathland, grassy sand dunes and open savannah woodland. Note the correlation between the estimated range of the coastal taipan and climate/vegetation zones.

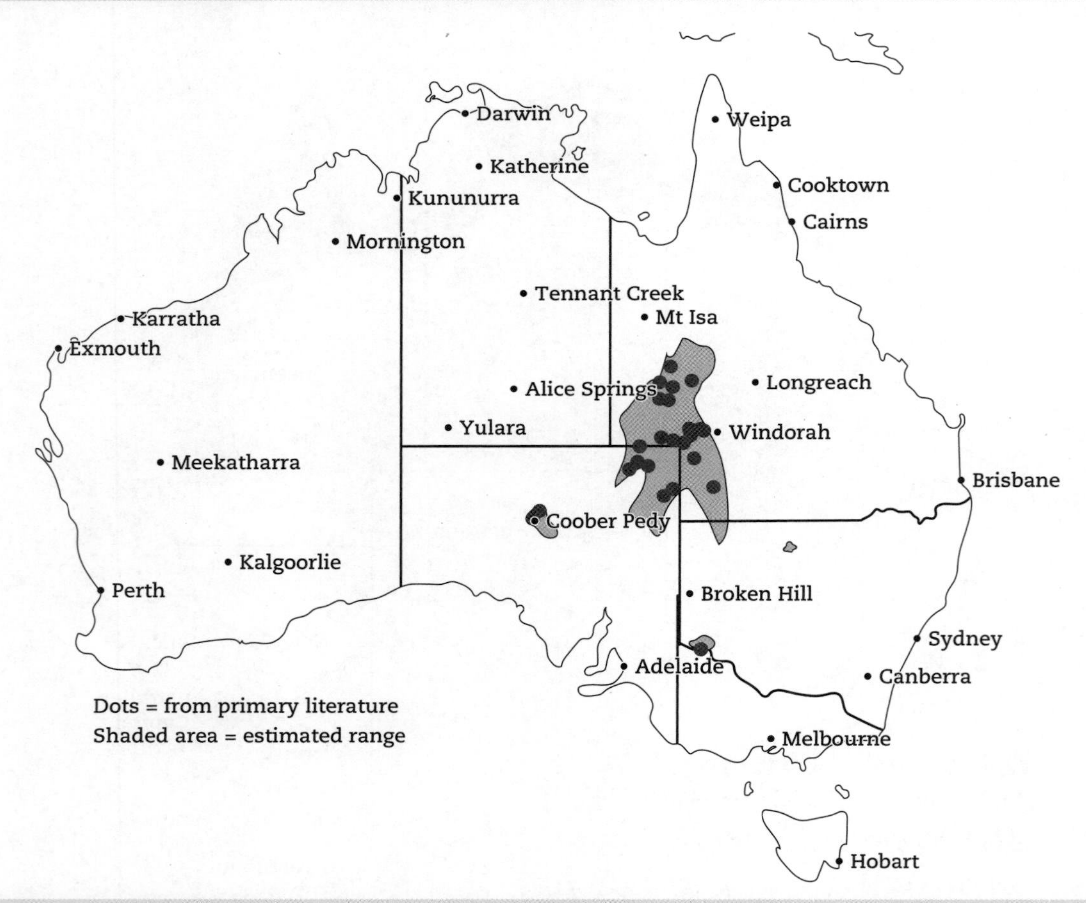

FIGURE 1.18 The inland taipan (*Oxyuranus microlepidotus*) occupies semi-arid habitats with low rainfall and sparse vegetation. It lives in deserts or plains with deep cracking-clays and cracking-loams of the floodplains. Note the correlation between the estimated range of the inland taipan and climate/vegetation zones.

This relationship is best summarised by identifying an organism's **tolerance limits** or the upper and lower limits of a particular environmental condition that allows the species to survive (Figure 1.19).

Hint
Think about what might limit the inland taipan from occupying more of Australia's semi-arid habitats.

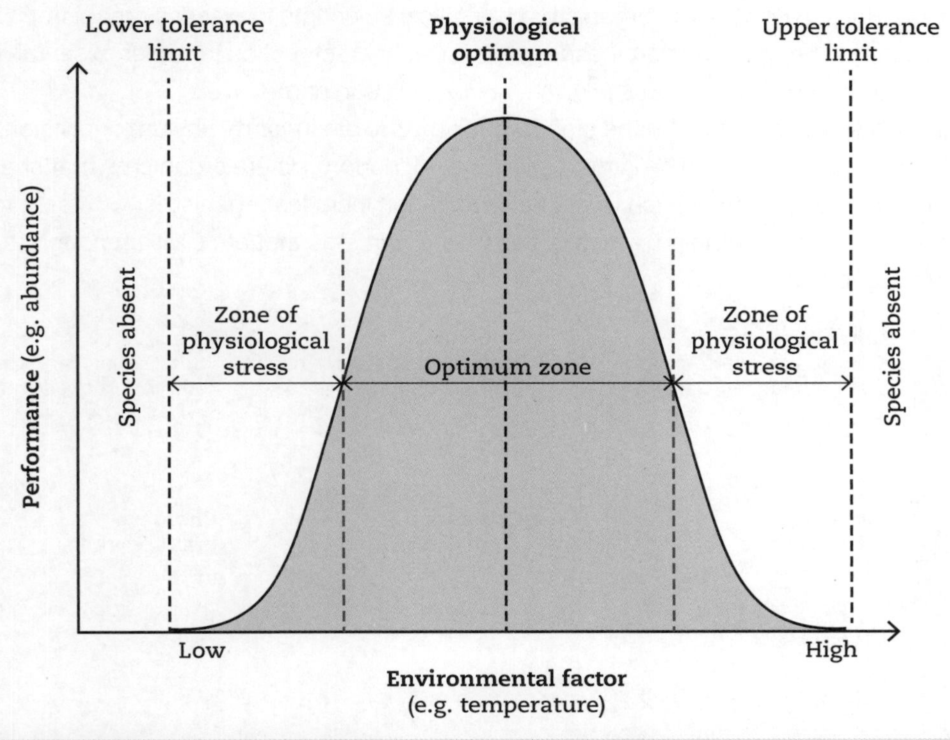

FIGURE 1.19 The effect of an environmental factor on the likely survival of an organism

Species with wide tolerance limits can survive and expand across increasing areas. An Australian example is the cane toad. In 1935, a total of 102 cane toads were brought to Australia to breed and 2400 were subsequently released around Ingham, Ayr, Mackay and Bundaberg. Their main requirement is constant access to water. Figure 1.20 shows the success of the cane toad expansion.

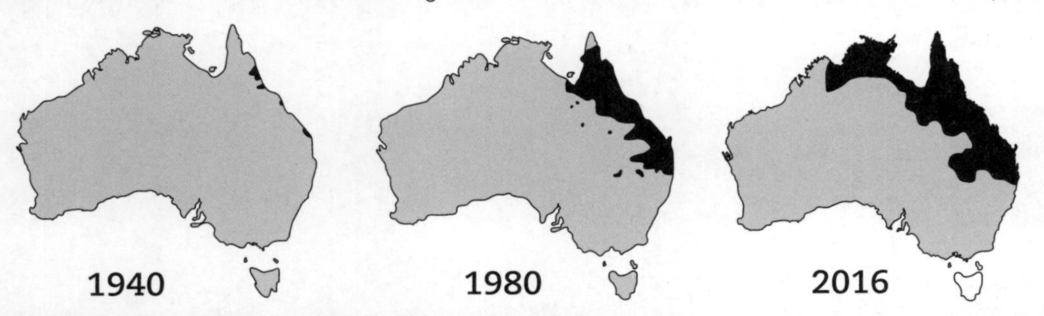

FIGURE 1.20 The expansion of cane toad populations between 1940 and 2016

1.2 Classification processes

Classification is a way of organising information from broad to specific categories. In biology, classification is based on examining and analysing the physical features or **morphology**, reproductive methods and molecular sequences across multiple species. Classification helps to understand key features and interactions of organisms. The method chosen for classification is determined by the data being collected or the question being answered.

1.2.1 Similarity of physical features: the Linnaean system

Carolus Linnaeus (1707–1778) is considered to be the founder of **taxonomy** and **binomial nomenclature** (rules used to name organisms) that allow scientists to discuss organisms without miscommunication. The groupings are based on physical characteristics (or other observable features) that are shared. Each species is given a two-part Latin name.

Linnaeus introduced the basis for the standard **hierarchy** of kingdom, phylum or division, class, order, family, genus and species. The largest grouping is kingdom, where organisms that share few similarities are grouped together. Each level of classification includes organisms with more traits in common up to the *species* grouping, which consists of organisms that are similar enough to produce fertile offspring together.

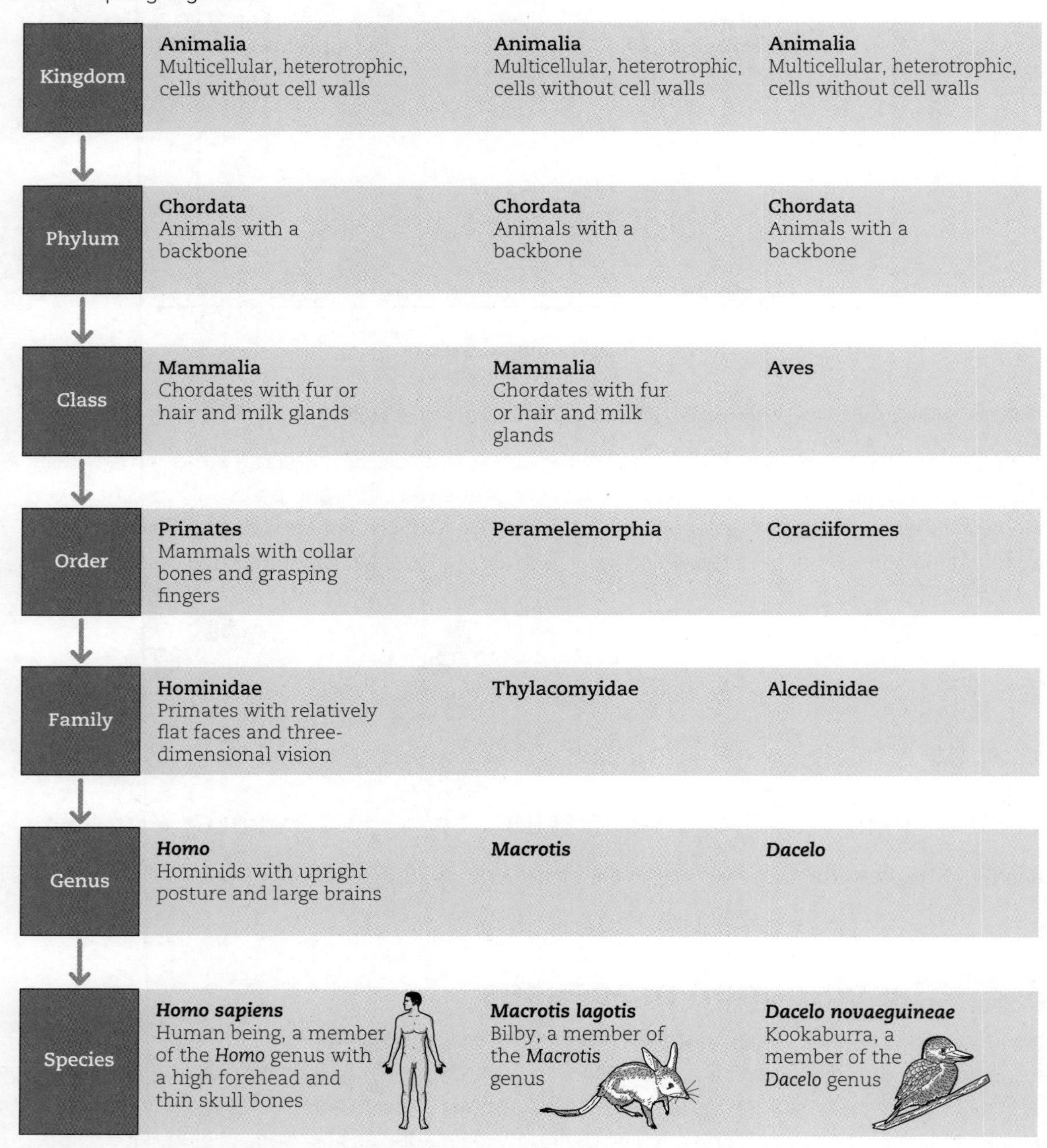

FIGURE 1.21 A comparison of the classification of humans, bilbies and kookaburras

Figure 1.21 compares the classification of humans, bilbies and kookaburras. Table 1.6 shows that kookaburras share only the ability to move and a backbone with humans, whereas bilbies share these features as well as hair/fur and milk production.

TABLE 1.6 A summary of similarities between human, bilby and kookaburra

Organism	Multicellular, heterotrophic, cells without cell walls	Backbone	Hair/fur and milk	Collarbones and grasping fingers	Flat face and 3D vision	Upright posture and large brain	Thin skull and high forehead
Human	✓	✓	✓	✓	✓	✓	✓
Bilby	✓	✓	✓	–	–	–	–
Kookaburra	✓	✓	–	–	–	–	–

This information can be turned into a simple dichotomous key in either of the formats shown in Figure 1.22. Note that each choice has only *two* options (the dichotomy).

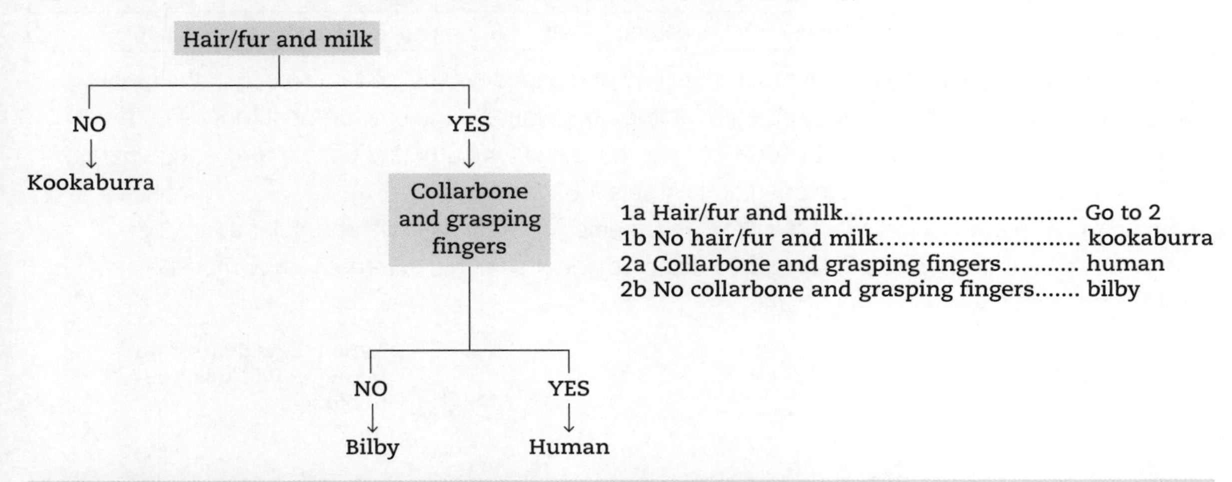

FIGURE 1.22 Two versions of a dichotomous key based on the morphologies listed in Table 1.6

Changes have been made to the Linnaean system as scientists have obtained new information about fossils, molecular biology and DNA sequencing.

Currently, most biologists agree that above the level of kingdom are the three domains Bacteria, Archaea and Eukaryota, as shown in Figure 1.23.

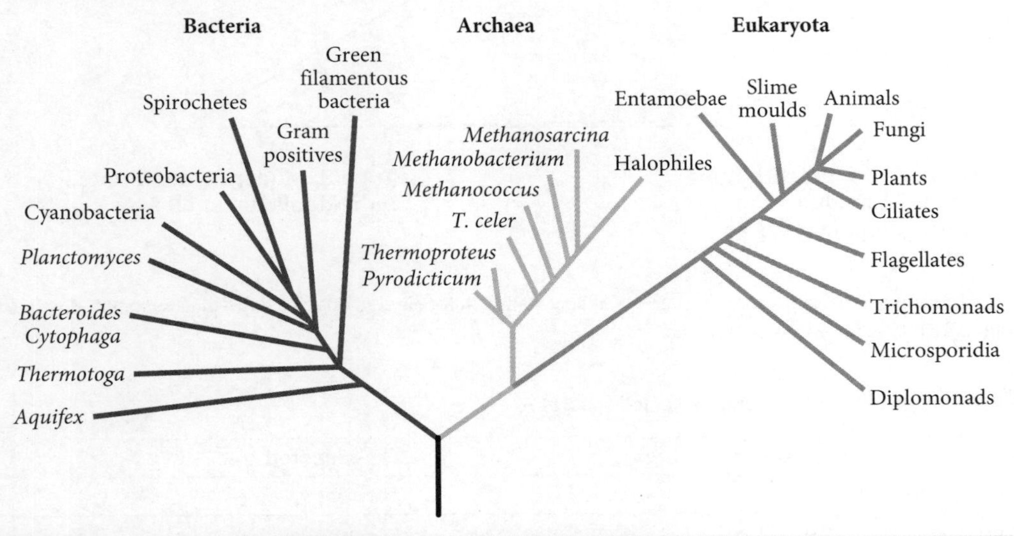

FIGURE 1.23 The currently accepted domains of life are Bacteria, Archaea and Eukaryota.

1.2.2 Methods of reproduction: sexual or asexual, K-selected or r-selected

Reproduction of organisms produces offspring. The type and number of offspring and level of parental involvement in raising the offspring vary significantly among species depending on whether they reproduce by **sexual reproduction** or **asexual reproduction** (Table 1.7).

TABLE 1.7 Attributes of sexual and asexual reproduction

Attribute	Sexual reproduction	Asexual reproduction
Parents	Requires male and female gametes	One parent
Genetic variation in offspring	Offspring genetically different from parent and each other	Offspring genetically identical to parent and each other
Cell division involved	Meiosis (see Unit 4 Topic 1)	Mitosis
Reproduction rate	Low	High
Parental care	Higher level of parental input	Lower level of parental input

Species whose populations vary around the carrying capacity (see Unit 3 Topic 2) in their environment are referred to as 'K-selected'. Like their populations, the environment of K-selected species also tends to be stable. r-selected species have highly fluctuating populations and generally live under environmentally variable conditions (Table 1.8).

Survivorship curves indicate how likely any species is to survive to a particular age. Most organisms are somewhere in between the r-selected and K-selected extremes, shown by the survivorship curve type II in Figure 1.24.

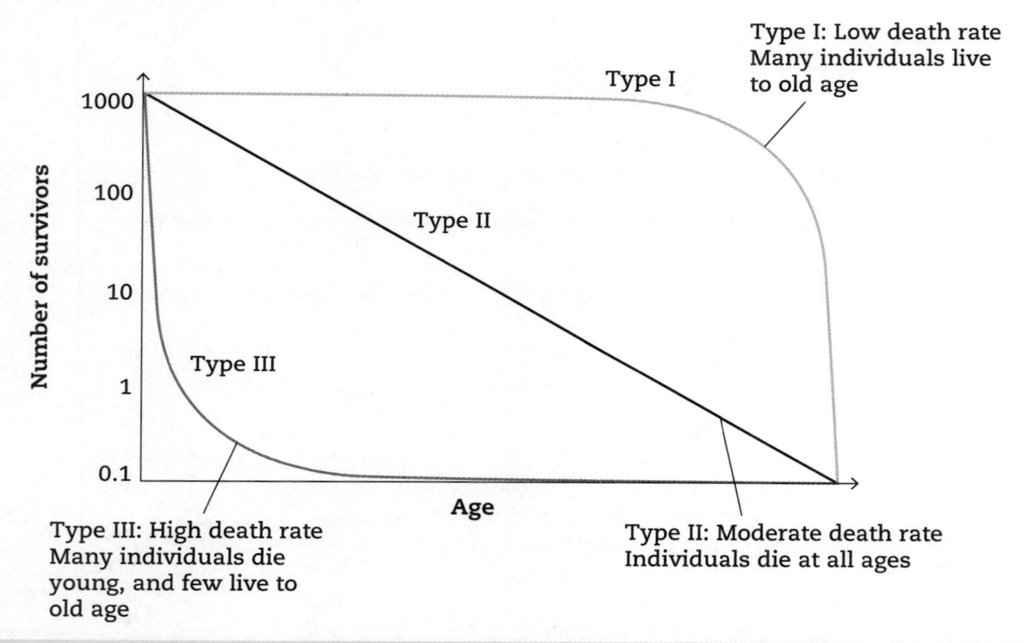

FIGURE 1.24 Typical survivorship curves. Type I is associated with K-selected species, type III is associated with r-selected species, and type II is somewhere in between.

TABLE 1.8 Attributes of K-selected and r-selected species

Attribute	K-selected	r-selected
Reproductive rate	Low	High
Reproductive age	Late	Early
Number of offspring	Small number	Large number
Parental care	High level of parental care input	Low level of parental care input
Age of maturation	Late maturity	Early maturity
Body size	Large	Small

››

Attribute	K-selected	r-selected
Life expectancy	Longer	Short
Survivorship	Type I (low mortality)	Type III (high mortality)
Environment	Stable	Unpredictable
Population	Stable	Variable
Australian examples	Wedge-tailed eagle iStock.com/Andrew Hayson	Mosquito Shutterstock.com/ Digital Images Studio

1.2.3 Molecular sequences: molecular phylogeny or cladistics

DNA and RNA sequences that occur repeatedly over generations and across species with little or no change are referred to as 'conserved'. Conserved sequences are assumed to acquire **mutations** at a constant rate. This means a time frame can be put on evolutionary relatedness (see also Unit 4 Topic 2).

An example is sequencing a short conserved section of mitochondrial DNA to produce a strip of nucleotide sequence resembling a barcode. The barcode can be compared to the barcodes of other species and analysed for evolutionary relatedness (Figure 1.25). **Molecular sequences** of amino acids and RNA nucleotides can also be analysed and interpreted to infer evolutionary relatedness.

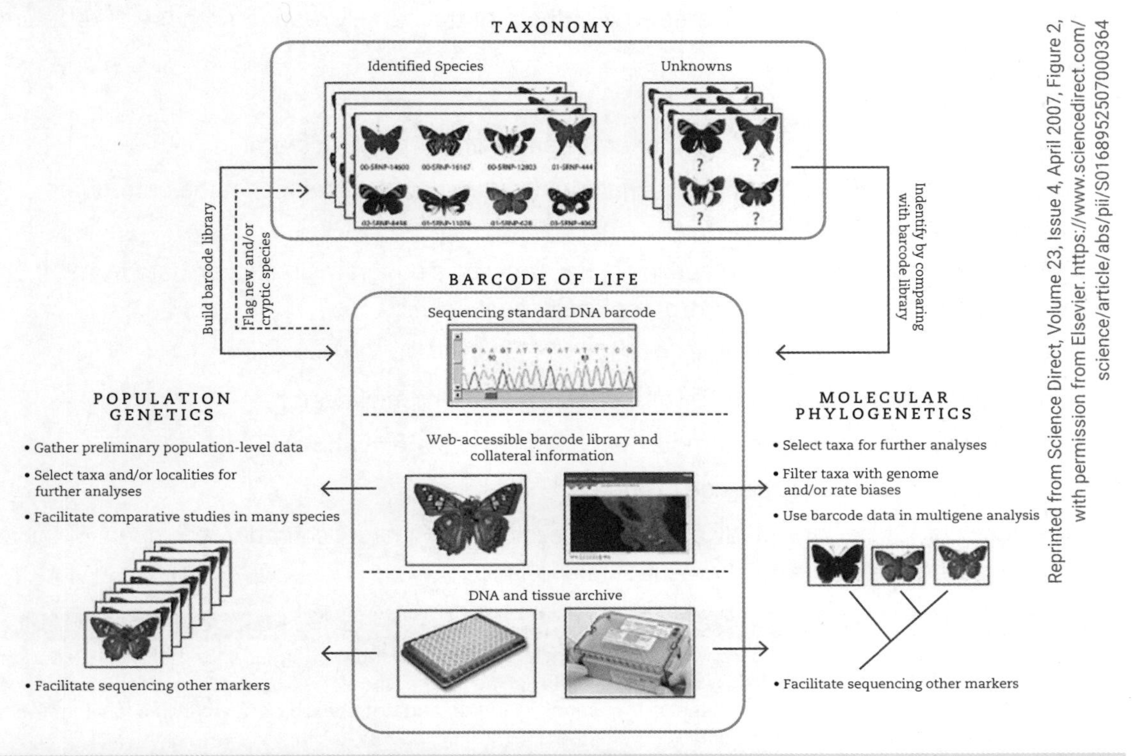

Reprinted from Science Direct, Volume 23, Issue 4, April 2007, Figure 2, with permission from Elsevier. https://www.sciencedirect.com/science/article/abs/pii/S0168952507000364

FIGURE 1.25 How molecular sequences, such as DNA barcodes, are used in taxonomy

Defining species

There are several ways to define 'species':

1 **Biological species concept**: a species is a group of organisms that can reproduce with one another in nature and produce fertile offspring. Thus, a horse and a donkey are different species because a mule (offspring of a female horse and male donkey) and hinny (offspring of a male horse and female donkey) are infertile. This is the most commonly used definition.
   - Problem 1: A female liger (male lion × female tiger) is fertile and can breed with other ligers, tigers and lions. Normally, tigers and lions would not meet so this definition of species assumes geographically isolated populations.
   - Problem 2: This definition does not allow for organisms that reproduce asexually.

   These problems can be addressed by other species concept definitions.
2 Ecological species concept: a species is a line of organisms kept separate by ecological forces. Stabilising selection maintains a species' integrity, while disruptive selection can lead to new species (see Unit 4 Topic 2).
3 Phylogenetic species concept: a species is a group of organisms that have a common ancestor, usually determined through genetic analysis.
4 Evolutionary species concept (can include asexual and extinct species): an evolutionary species

> 'is a single lineage of ancestor–descendant populations of organisms which maintains its identity from other such lineages [in space and time] and which has its own evolutionary tendencies and historical fate'.

Cladistics

Whichever species definition is used, a **clade** is a group of organisms showing a single ancestor group and the descendant groups (Figure 1.26):

- Individuals are more closely related to members of the same group than those of different groups.
- Branching points are called nodes.
- Groups being investigated (taxa) occur at the endpoint of the cladogram.
- The species with the least number of characteristics in common represent the outgroup (establishes baseline properties).
- These relationships are represented in diagrams called cladograms based on data from morphological characteristics and/or molecular sequences.

There are three basic assumptions in cladistics:

1 Groups of organisms are *related by descent* from a common ancestor.
2 The branching pattern is dichotomous (each branch is a **bifurcation**).
3 Characteristics of lineages change over time.

Therefore, cladistics is the study of common evolutionary history shared by members of the same taxonomic group. Clades are a statement of relationship.

Hint

The terms 'cladogram', 'phylogenetic tree' and 'phylogram' are often used interchangeably. Sometimes 'cladogram' predicts the possible evolutionary history of a group, while 'phylogenies' show the true evolutionary history. At other times, 'cladogram' suggests that the lengths of the branches in the diagram are arbitrary, while in a 'phylogeny', the branch lengths indicate the amount of character change.

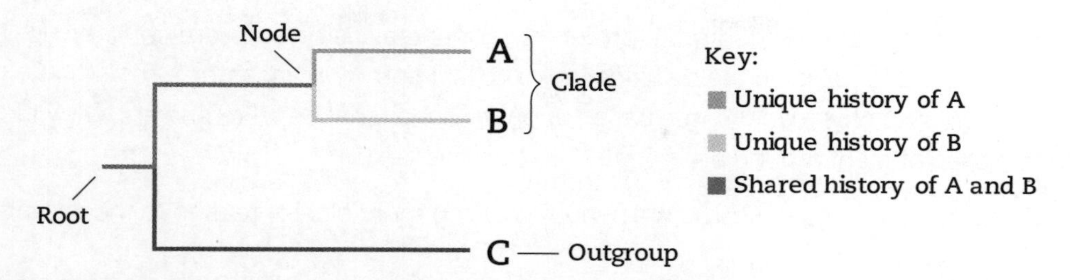

FIGURE 1.26 The parts of a cladogram include node, root, clade and outgroup.

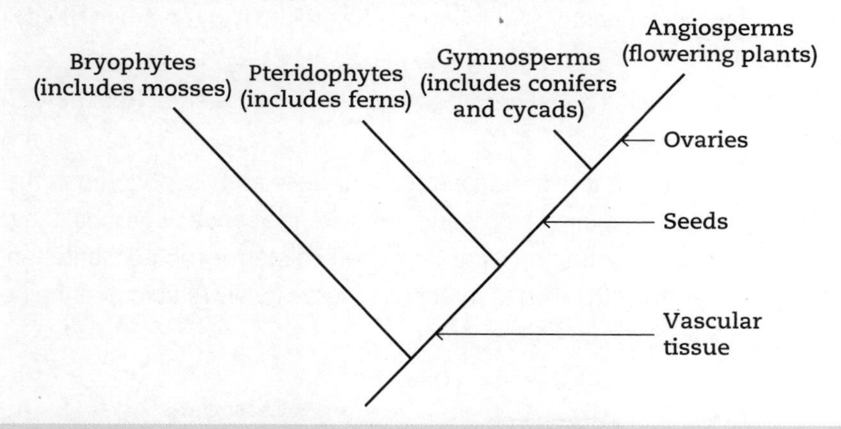

FIGURE 1.27 At each branching point, some of the offspring of the ancestor developed a new feature, while others did not. This evolution of features has resulted in a wide, but interrelated, variety of life on Earth.

In time-rooted cladograms, all groups are generally lined up. All groups are usually still living (extant). Sometimes there will be a time scale. The focus is on the *most recent* **common ancestor**, indicated by a node. Figure 1.28 shows a phylogenetic tree and two cladograms.

Question 1: In Figure 1.28a, are turtles more closely related to amphibians or mammals?

1 Highlight the groups of interest: turtles, amphibians and mammals.

2 Redraw the cladogram with the groups of interest. See Figure 1.28b for two options for redrawing the cladogram.

3 Determine which has the most recent common ancestor.

Answer: The most recent common ancestor occurs between mammals and turtles, not mammals and amphibians; therefore, turtles are more closely related to mammals.

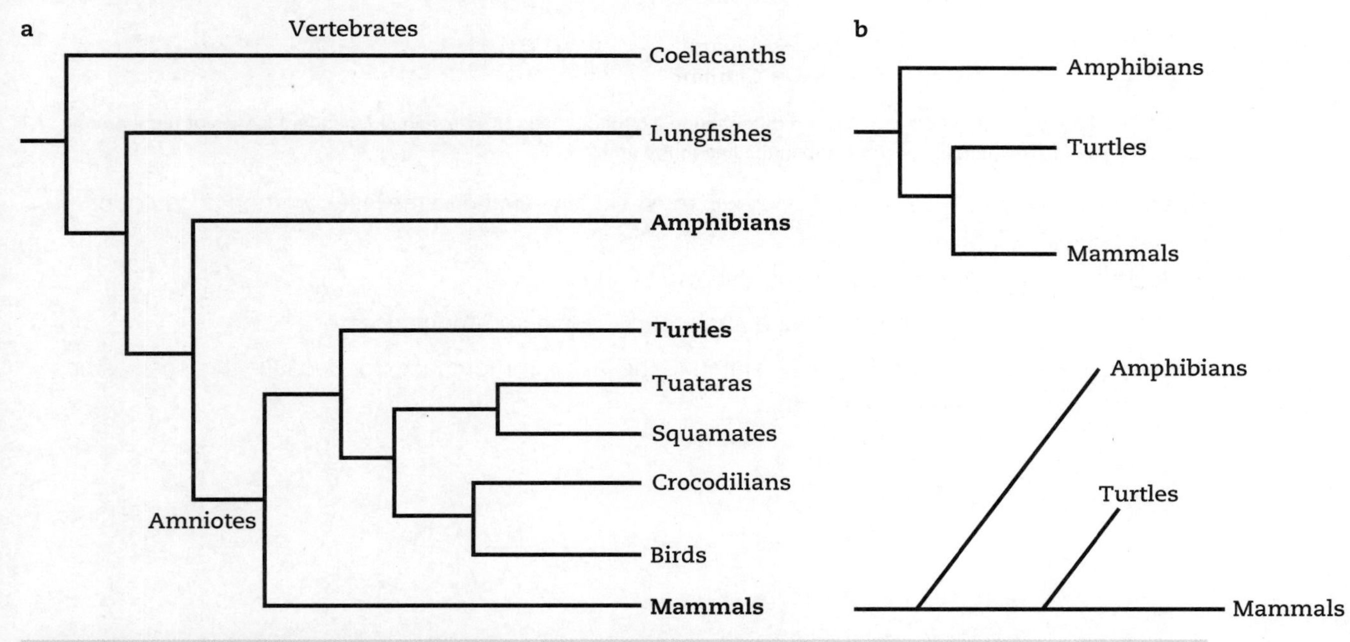

FIGURE 1.28 a A phylogenetic tree of bird species and all animals. b Two options for redrawing the cladograms, including only the species of interest.

9780170459136

U3 – TOPIC 1

In genetic distance-rooted cladograms (phylograms), generally all groups are not lined up. They may have a scale. The focus is on the *distance* from the last common ancestor.

Figure 1.29 shows a phylogenetic tree based upon mitochondrial genetic markers, showing independent losses of flight among birds.

Question 2: According to Figure 1.29, is the penguin or emu more closely related to the brush turkey?

1 Highlight the groups of interest.
2 Consider only branch length.
3 Determine the last common ancestor of the two species relative to the brush turkey (shown by arrow).
4 Highlight the distance between the last common ancestor and the emu. Repeat for the penguin (shown by bold lines).

Answer: The branch between the last common ancestor and the emu is longer than the branch of the last common ancestor and the penguin. The greater the distance between groups, the more distantly related they are. This is because the branch length in genetic distance-rooted cladograms represents the number of changes. Therefore, the shorter branch is more closely related and the penguin is the closer relative.

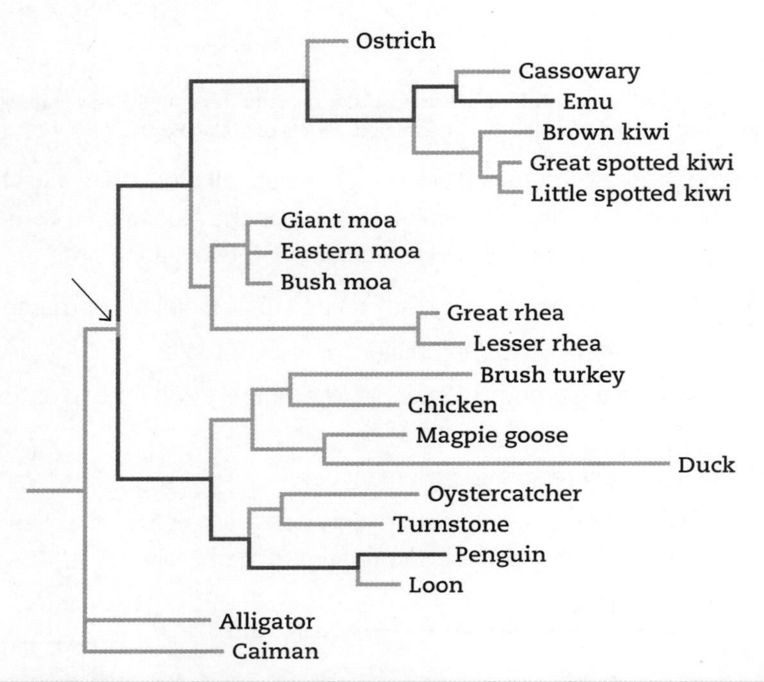

FIGURE 1.29 A phylogenetic tree based on mitochondrial genetic markers, showing independent losses of flight among birds. The line to the penguin is shorter than the line to the emu.

Question 3: In Figure 1.29, which species is most closely related to the last common ancestor of these species?

1 Identify the last common ancestor – shown by the arrow.
2 Find the species with the shortest branch from the last common ancestor.

Answer: The alligator is most closely related to the last common ancestor because this species has the shortest branch (see Figure 1.30).

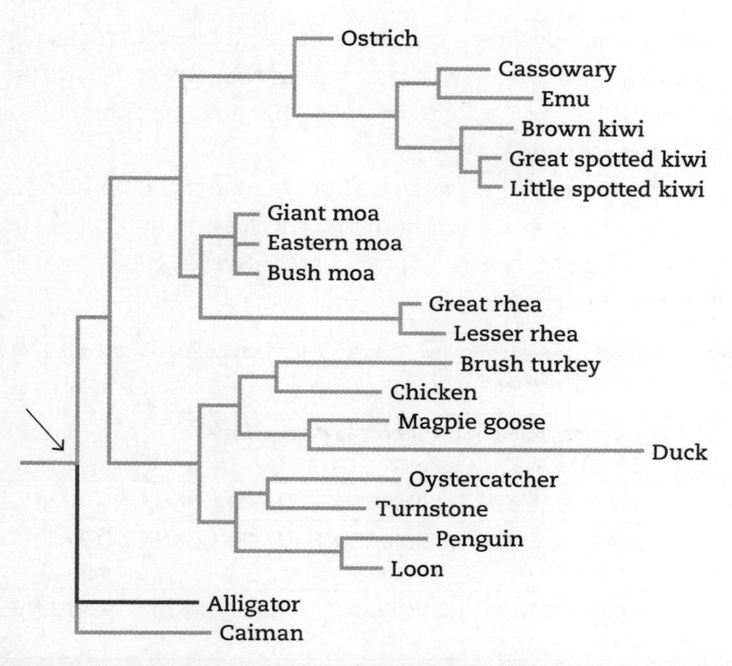

FIGURE 1.30 In the phylogenetic tree, the line from the common ancestor to the alligator is the shortest line.

How to construct a cladogram

In Table 1.6 of morphologies from section 1.2.1 (page 17), you can see that each characteristic is represented by a node on the cladogram.

TABLE 1.9 Characteristics represented by nodes

Organism	Move on their own	Backbone	Hair/fur and milk	Collarbones and grasping fingers
Human	✓	✓	✓	✓
Bilby	✓	✓	✓	–
Kookaburra	✓	✓	–	–

Steps for cladogram construction:

1 Begin with the characteristic(s) that all organisms have (in this example: move on their own and backbone).
2 Draw the first branch. On the first branch, write the name of the organism that does not share the next characteristics (in this example: kookaburra).
3 After the first branch, write the two characteristics that the remaining organisms share (in this example: hair/fur and milk).
4 Draw the second branch. Again, write the name of the organism that does not share the next characteristics (in this example: bilby).
5 After the second branch, write the characteristics the remaining organisms have (in this example: collarbone and grasping fingers).
6 The third branch has the name of the remaining organism (in this example: human).

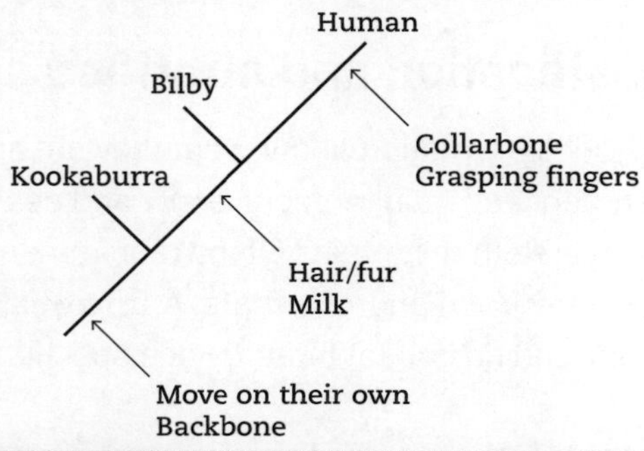

FIGURE 1.31 A cladogram based on the information in Table 1.9

Cladograms (or phylogenetic trees) can also be constructed by using molecular sequences of nucleotides or amino acids. Table 1.10 shows an amino acid sequence of cytochrome c from five different species. Cytochrome c is a small protein (about 100 amino acids) that is important in the successful functioning of mitochondria.

The least number of differences in the sequence from human cytochrome c suggests the closest relationship between that species and humans. This information can be used to construct a cladrogram, similar to the series of steps listed previously. However, with amino acid sequences, begin with the species that has the most *differences*.

TABLE 1.10 Cytochrome c amino acid sequence for five species. Each letter represents a different amino acid and differences in amino acids are highlighted in bold

Organism	Amino acids in cytochrome c sequence	Number of differences
Human (*Homo sapiens*)	MG–DVEKGK KIFIMKCSQC HTVEKGGKHK TGPNLHGLFG	0
Chimpanzee (*Pan troglodytes*)	MG–DVEKGK KIFIMKCSQC HTVEKGGKHK TGPNLHGLFG	0
Emu (*Dromaius novaehollandiae*)	MG–D**I**EKGK KIF**VQ**KCSQC HTVEKGGKHK TGPNL**N**GLFG	4
Pacific lamprey (*Lampetra tridentata*)	MG–DVEKGK K**V**F**VQ**KCSQC HTVEK**A**GKHK TGPNL**S**GLFG	5
Fruit fly (*Drosophila melanogaster*)	MG**SG**D**A**E**N**GK KIF**VQ**KC**A**QC H**TY**E**V**GGKHK **V**GPNL**G**G**VV**G	14

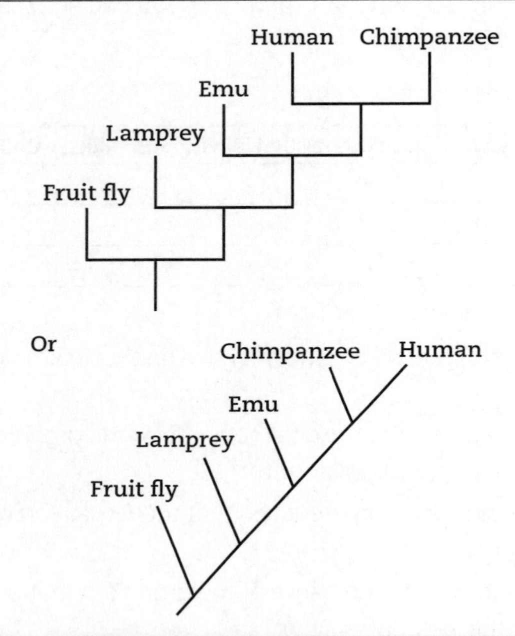

FIGURE 1.32 Cladograms based on the number of differences in the amino acid sequence of cytochrome c

Note: If the mutation rate is known, then the length of the branch can be scaled properly.

Organisms can also be classified according to species interactions (see section 1.1.2).

Hint
A microhabitat differs from a niche because a niche includes the habitat *and* the organism's role in the environment.

1.2.4 Ecosystem classification and stratified sampling

Ecosystems are classified to provide information about what they are and how they function. This helps to establish laws to protect habitats, monitor biodiversity and describe habitat requirements.

Remember that a **habitat** is an area that contains all the requirements for an organism to survive and reproduce, including shelter, water, food and/or a mate. A **microhabitat** is a smaller part of that area that differs from the surrounding habitat but provides specific physical conditions in the immediate vicinity of the organism. An **ecoregion** is a large area that has biogeographical features that are similar, including ecosystem interactions, environmental conditions and native species. There are transition areas that link each ecoregion, so the boundaries are not fixed. Ecoregions may be terrestrial or aquatic.

There are four ecosystem classification systems suggested in the syllabus.

1 Specht's classification (1970) was developed in Australia and gives information about the height of the dominant vegetation **stratum** and its percentage of **canopy** cover. This information is collected by sampling, ignoring emergent trees.

FIGURE 1.33 This canopy is 10–30 metres high with a 45–50% canopy cover (during December); therefore, this is an open forest.

TABLE 1.10 Structural classification of vegetation based on Specht's system

Life form of tallest stratum	Projective foliage cover of the tallest stratum				
	>70%	51–70%	31–50%	10–30%	<10%
Trees >30 m	Tall closed forest	Tall forest	Tall open forest	Tall woodland	n/a
Trees 10–30 m	Closed forest	Forest	Open forest	Woodland	Open woodland
Trees <30 m	Low closed forest	Low forest	Low open forest	Low woodland	Low open woodland
Shrubs >2 m	Closed scrub	Scrub	Open scrub	Tall shrubland	Tall open shrubland
Shrubs (S) 0.25–2 m	Closed heathland	Heathland	Open heathland	Shrubland	Open shrubland
Shrubs (NS) 0.25–2 m	n/a	n/a	Low shrubland	Low shrubland	Low open shrubland
Shrubs (S) <0.25 m	n/a	n/a	n/a	Dwarf open heathland	Dwarf open heathland
Shrubs (NS) <0.25 m	n/a	n/a	n/a	Dwarf shrubland	Dwarf open shrubland
Hummock grasses	n/a	n/a	n/a	Hummock grassland	Open hummock grassland
Tussock grasses	Closed grassland	Grassland	Grassland	Open grassland	Very open grassland
Sedges	Closed sedgeland	Sedgeland	Sedgeland	Open sedgeland	Very open sedgeland
Herbs	Closed herbland	Herbland	Herbland	Open herbland	Very open herbland
Ferns	Closed fernland	Fernland	Fernland	n/a	na

Note that this classification gives no information about the species or abiotic factors present in the sample area. It does not take into account the deciduous nature of some trees.

S, sclerophyll (hard leaf); NS, non-sclerophyll; n/a, does not occur naturally.

Trees: woody plant more than 5 m tall, usually with a single stem.

Shrubs: woody plant less than 8 m tall, frequently with many stems arising at or near the base.

Hint

Make sure you measure the *canopy cover only*, not the cover provided by all the strata. You will need to consider the time of year because some trees are deciduous.

Figure 1.34 shows that the tallest stratum is trees (woody plant, <10 metres, single trunk), with a percentage cover of 5–9%. This makes it a low woodland.

FIGURE 1.34 Low woodland in south-east Queensland

2 The Holdridge life zones refer to a system for classifying land areas on the basis of potential evapotranspiration, biotemperature and precipitation. Lines are ruled across the grid based on the data collected, as shown in Figure 1.35:

- one line for the potential evapotranspiration ratio (in this example: 1.75)
- one line for the average total annual precipitation (in this example: approximately 850 mm)
- one line for the biotemperature (in this example: 15°C).

The intersection point of the three lines identifies the life zone. In the example in Figure 1.35, the three lines intersect at 'dry forest'.

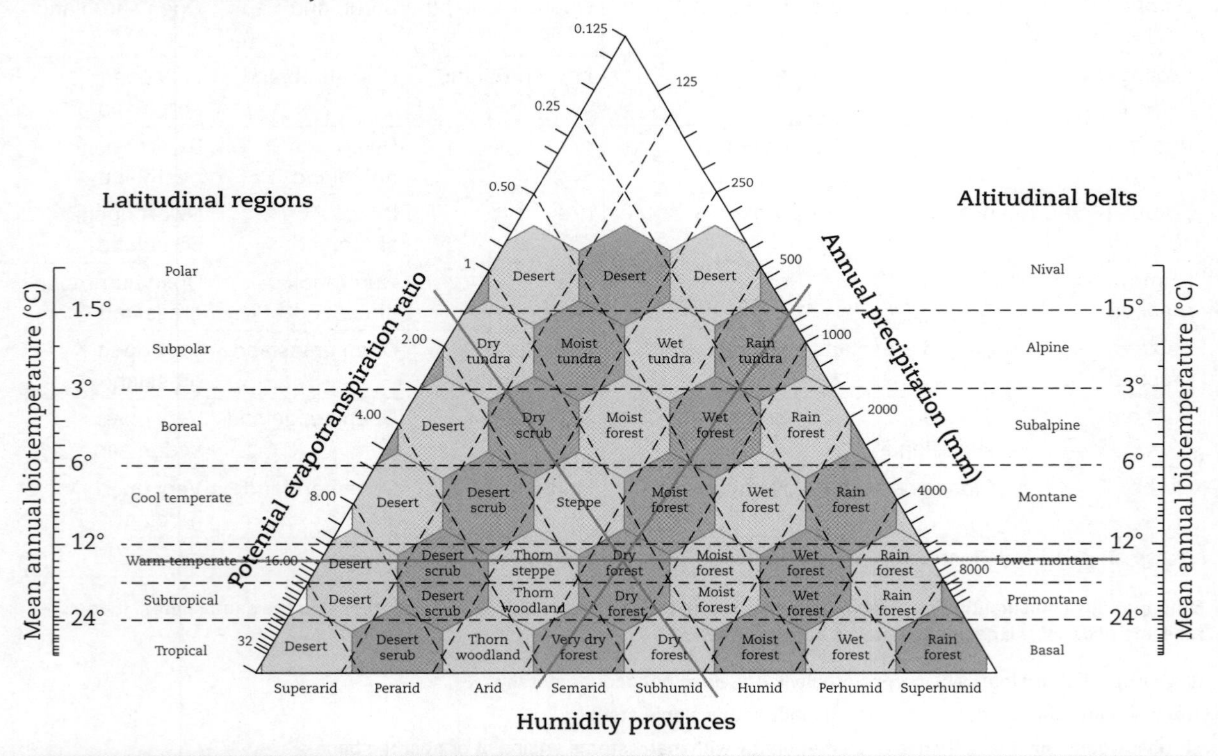

FIGURE 1.35 The Holdridge life zone classification system

3 The Australian National Aquatic Ecosystem (ANAE) consists of three levels. Levels 1 and 2 are large-scale measures. Level 3 is specific to types of aquatic ecosystems:

- Level 1: Regional scale
- Level 2: Landscape scale
- Level 3: Aquatic classes, systems and habitats.

ANAE structure											
Level 1		Regional scale (Attributes: hydrological, climate, landform)									
Level 2		Landscape scale (Attributes: water influence, landform, topography, climate)									
Level 3	Class	Surface water						Subterranean			
	System	Marine	Estuarine	Lacustrine	Palustrine	Riverine	Floodplain	Fractured	Porous sedimentary rock	Unconsolidated	Cave/karst
	Habitat	Pool of attributes to determine aquatic habitats (e.g. water type, vegetation, substrate, porosity, water source)									

FIGURE 1.36 Levels used when classifying an aquatic ecosystem according to the ANAE system

4 The European Nature Information Service (EUNIS) habitat classification is a European system for habitat identification. The classification is hierarchical and covers all types of habitats from natural to artificial, from terrestrial to freshwater and marine. The habitat types are identified by specific codes, names and descriptions.

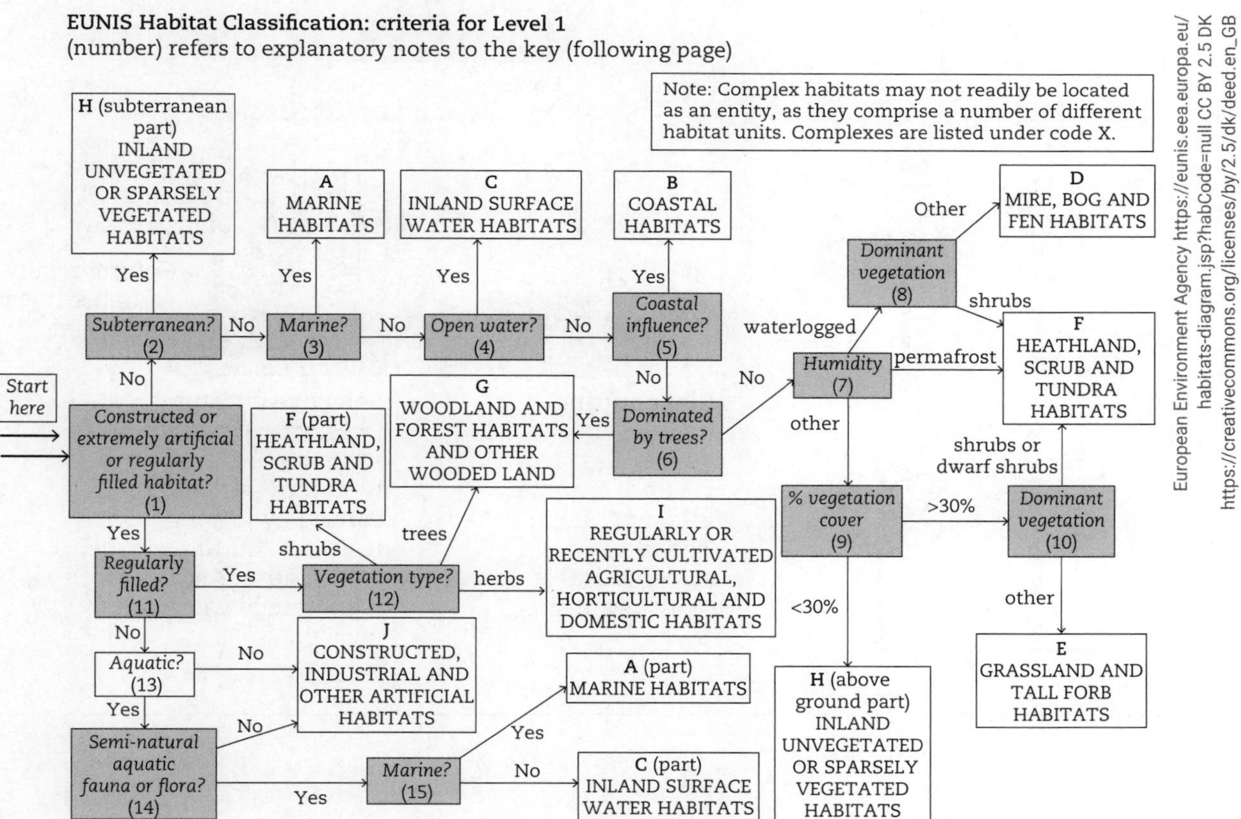

FIGURE 1.37 Level 1 criteria for EUNIS classification. This database is now online.

1.2.5 Ecosystem management

Once a system is classified on the basis of data collected, ecologists can compare expected numbers, types and distribution of species and observe changes over time. Ecologists can decide on whether to introduce management strategies to conserve the composition, structure and function of an ecosystem.

Ecosystem functions include primary production, decomposition and nutrient recycling. **Ecosystem services** include being a source of food, water and oxygen and regulating climate.

The aims of ecosystem management are to protect and sustain the functions and services of the ecosystem.

TABLE 1.11 Summary of the importance of ecosystems and possible management strategies

Ecosystem	Why it is important	How to manage it
Old growth forest A stable but diverse structured forest with a mature canopy and multiple vegetation layers and the presence of hollows, large logs and gaps in the canopies due to fallen trees. The effects of past disturbances are now negligible.	• Help maintain the functions and services listed above as well as biodiversity • Used for foraging, basking, nesting. Higher levels of food resources. Deep litter layer and many hollows • Carbon storage	• Protection of old growth forests • Provision for forests with old growth attributes • Controlled logging using selective cutting

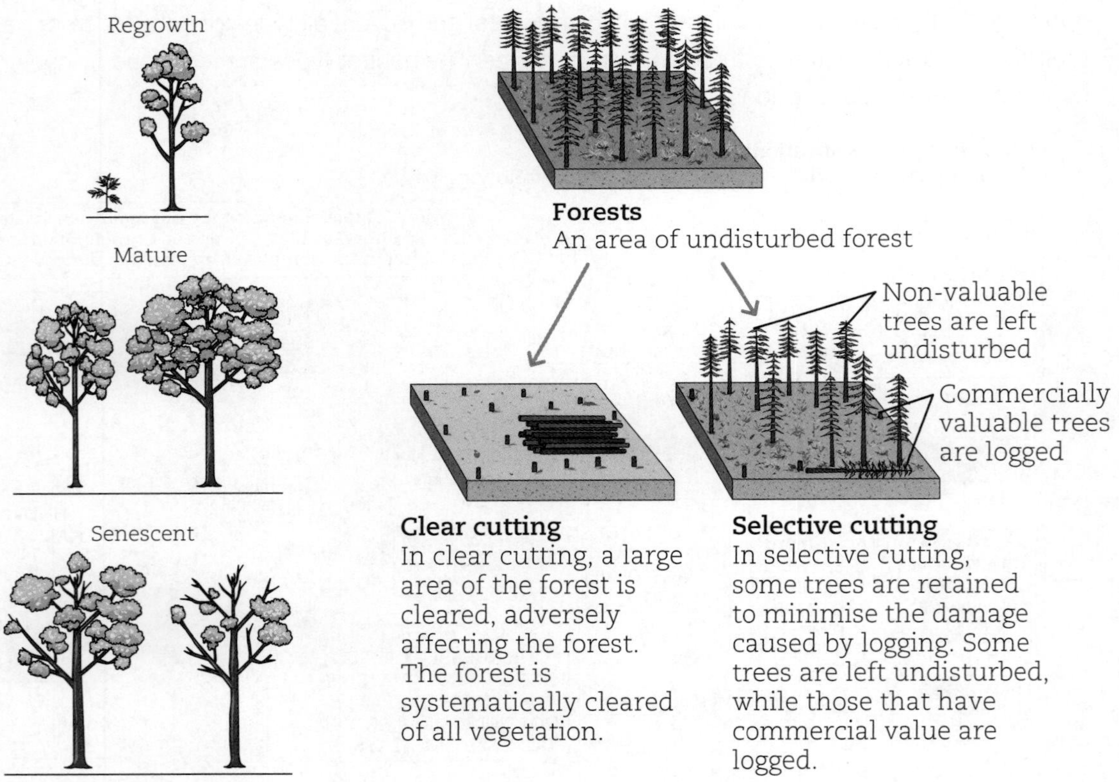

Regrowth

Mature

Senescent

Forests
An area of undisturbed forest

Non-valuable trees are left undisturbed

Commercially valuable trees are logged

Clear cutting
In clear cutting, a large area of the forest is cleared, adversely affecting the forest. The forest is systematically cleared of all vegetation.

Selective cutting
In selective cutting, some trees are retained to minimise the damage caused by logging. Some trees are left undisturbed, while those that have commercial value are logged.

››

Ecosystem	Why it is important	How to manage it
Soil The greatest soil biodiversity is found in natural ecosystems. Productivity is the capacity to support plant growth. Important functions affecting productivity within any soil are nutrient availability, air availability and physical support and essential nutrients.	To meet the nutrient and growth requirements of plants to achieve: • sustainable ecosystems • crop productivity.	Encourage soil biodiversity through: • crop rotation, irrigation, drainage • revegetation • re-establishing symbiotic relationships between plants and soil microbes • C:N:P ratios
Coral reef Skeletons of marine invertebrates form large underwater structures that provide 25% of all known marine species with food, shelter and breeding sites.	• Protect shoreline from erosion • Source of food and medicines • Provide nurseries for marine species	• Create marine protection areas • Reduce threats (e.g. invasive species) and control recreational usage • Minimise the severity of bleaching • Manage fishing

1.2.6 Stratified sampling

Strata can be vertical or horizontal, as shown in Figures 1.38–1.40.

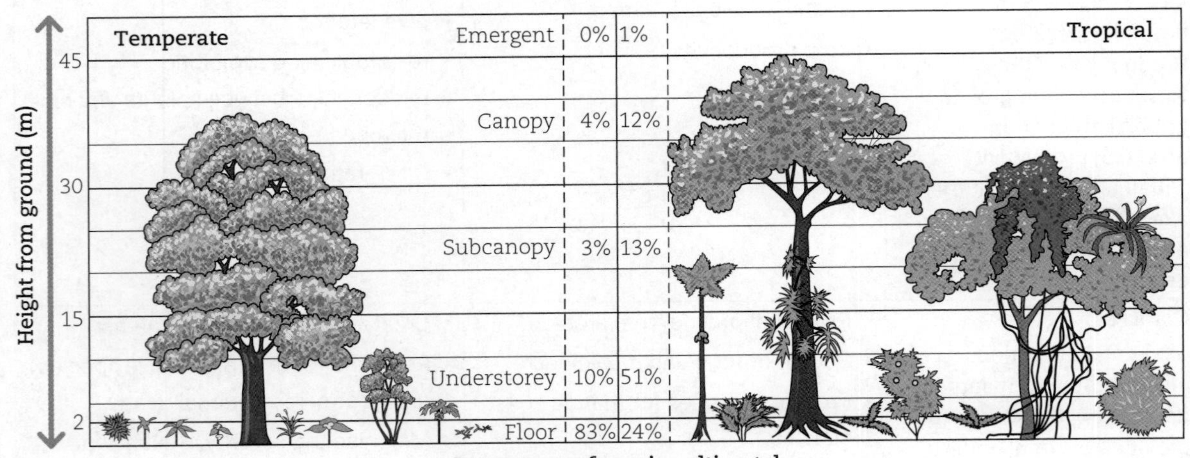

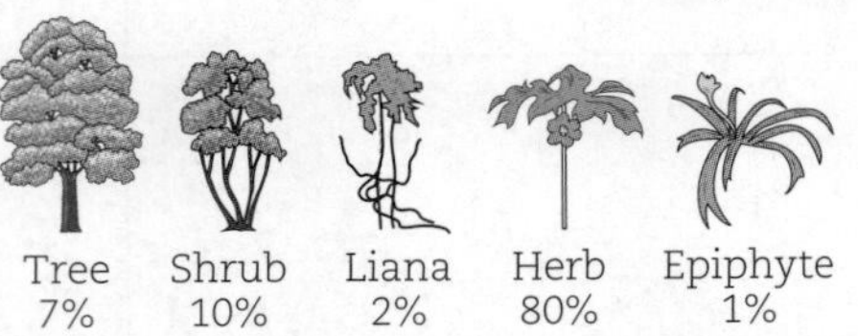

FIGURE 1.38 A comparison of vertical strata in temperate and tropical forests

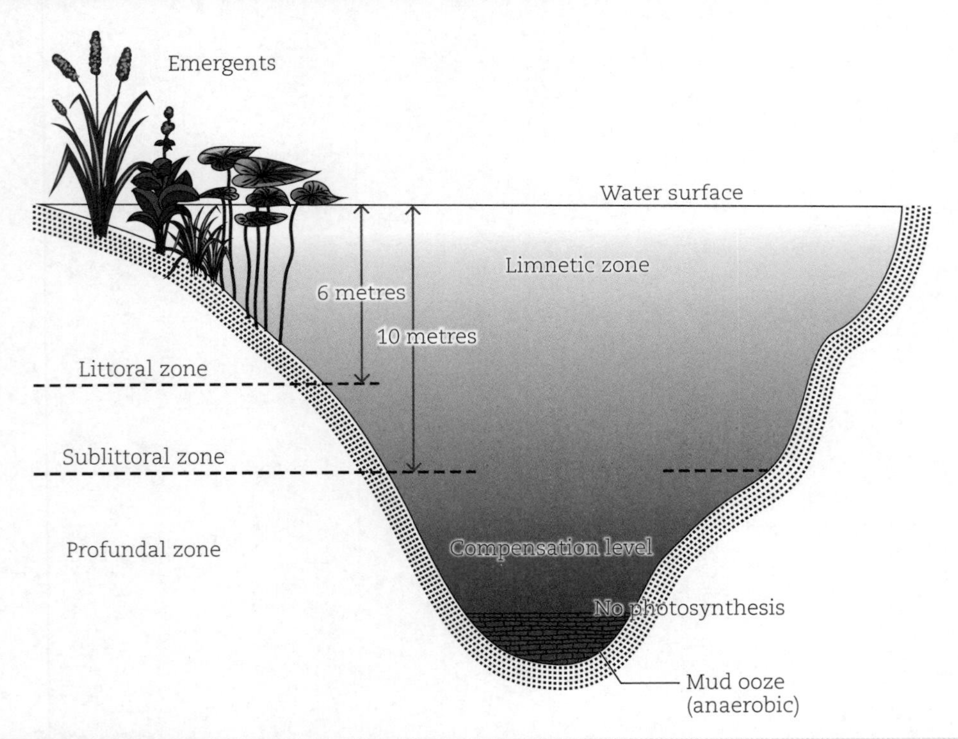

FIGURE 1.39 Stratification of a lake

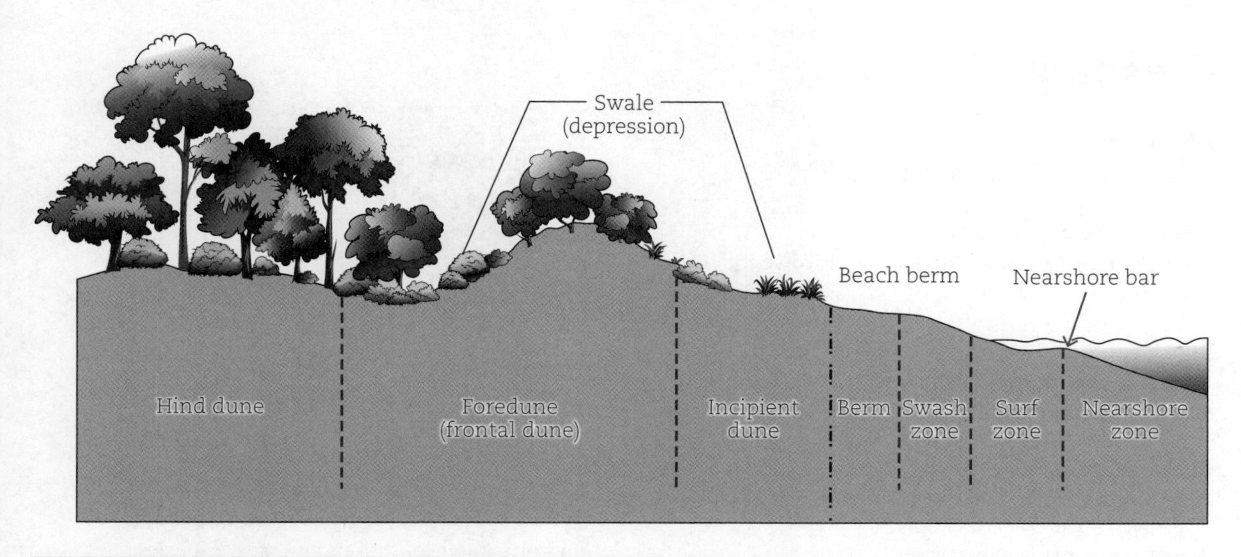

FIGURE 1.40 Horizontal stratification during the process of succession in a coastal dune system

Stratified sampling is used when the study area can be divided into defined, relatively **homogeneous** strata that are not overlapping. Sampling is then conducted separately in each stratum.

TABLE 1.12 Process of stratified sampling

Purpose – what are you trying to find out?	Estimating: density, population size, distribution of species Identifying: environmental gradients and profiles, zonation and stratification
Site selection	Choose homogeneous layers suited to the purpose of the study.
Survey technique	Transect or quadrat? (See Figure 1.7, page 8)
Bias minimisation	Bias occurs when the data does not reflect the actual situation. Minimise bias by: • increasing the sample size • choosing appropriate size of quadrat/length of transect • using random number generators (if random sampling).
	• using consistent counting criteria (e.g. does an organism have to be entirely in a quadrat?) and technique • calibrating equipment appropriately • noting equipment precision, e.g. grid quadrat versus forest densiometer.
Data presentation – quantitative	Quantitative data is initially recorded in tables, then presented in graphs, kite diagrams or transect profiles, depending on the purpose of the sampling.
Data presentation – qualitative	Descriptions and photographs can be organised in tables.

9780170459136

Glossary

A+ DIGITAL FLASHCARDS
Revise this topic's key terms and concepts by scanning the QR code or typing the URL into your browser.

https://get.ga/aplus-qce-bio-u34

Abiotic factor
A non-living part of the environment.

Asexual reproduction
A type of reproduction that does not involve the fusion of gametes.

Belt transect
A strip marked across a habitat within which distribution (and abundance) of species is recorded.

Bifurcation
When something divides into two branches or parts.

Binomial nomenclature
A naming system in which each species name has two parts; the first is the genus, the second is the species

Biodiversity
A measure of the variety of species (plants, animals and micro-organisms) in a given ecosystem.

Biological species concept
A group of organisms who can interbreed successfully and produce fertile offspring.

Biotic factor
A living part of the environment.

Canopy
A collection of individual plant crowns that form the above ground portion of plants.

Clade
A group of organisms that consists of a common ancestor and all its lineal descendants.

Classification
The scientific process of grouping organisms according to how similar they are.

Commensalism
A type of symbiosis where only one species benefits and the other does not benefit nor is harmed. + 0

Common ancestor
A population of individuals from which more than one species can trace descent by modification.

Competition
An interaction between different organisms or species for limited resources, usually to the detriment of one.

Competitive exclusion
Two species cannot occupy the same niche indefinitely.

Disease
An abnormal state of an organism that interferes with normal function, often caused by a pathogen.

Distribution
A geographic area in which a species is found.

Ecoregion
A large area that has biogeographical features that are similar, including ecosystem interactions, environmental conditions and native species.

Ecosystem functions
The sum of the activities of organisms (e.g. feeding, excreting, moving) and the effects these have on their environment.

Ecosystem services
The benefits gained by humans from their environment, including physical, biological and cultural benefits.

Habitat
An area that contains all an organism requires for survival and reproduction, including shelter, water, food and/or a mate.

Hierarchy
The classification of organisms into levels of increasing similarity, e.g. from kingdom to species.

Homogeneous
In ecology, indicates a lack of species diversity.

Interspecific interaction
An interaction that occurs between individuals of different species.

Intraspecific interaction
An interaction that occurs between individuals of the same species.

Line transect
A line, usually a string, across a habitat or part of a habitat, along which numbers of species can be recorded at regular intervals.

Microhabitat
A smaller part of an area that differs from the surrounding habitat but provides specific physical conditions in the immediate vicinity of the organism.

Molecular sequence
A continuous molecule of DNA, RNA or amino acids.

Morphology
The form and structure of living things.

Mutation
A change in the DNA sequence of an organism.

Mutualism
A type of symbiosis in which both species benefit. + +

Parasitism
A type of symbiosis in which one species benefits and the other is harmed. + –

Percentage cover
The percentage of area covered by each species in a sample area.

Percentage frequency
A measure of how often each species occurs in a sample area.

Predation
The killing of one organism by another for food.

Quadrat
A frame, usually square, used to select a sample area in a larger study area.

Random sampling
A type of sampling in which each sample has an equal chance of being chosen, i.e. unbiased.

Sexual reproduction
A type of reproduction that involves the fusion of gametes.

Species diversity
The number of species and abundance of each species in a given ecosystem.

Species evenness
The relative number of each species in a given ecosystem.

Species richness
The number of species in a given ecosystem.

Stratified sampling
A type of sampling in which a sample is taken of each stratum of a population.

Stratum
A vertical or horizontal layer of habitat within a larger ecosystem (plural: strata).

Survivorship curve
A graph showing the number of individuals in a population expected to survive to a given age.

Symbiosis
An interaction between individuals of two different species from which one or both benefits.

Systematic sampling
A type of sampling in which each sample taken represents an equal portion of the whole population.

Taxonomy
The science of naming and describing all living things.

Tolerance limits
The range of environmental factors (e.g. light, temperature) within which an organism can exist.

Revision summary

Use the following summary of syllabus dot points and key knowledge within Unit 3 Topic 1 to ensure that you have thoroughly reviewed the content. Provide a brief definition or comment for each item to demonstrate your understanding or code them using the traffic light system – green (all good), amber (needs some review), red (priority area to review).

Biodiversity	
• recognise that biodiversity includes the diversity of species and ecosystems	
• determine diversity of species using measures such as species richness, evenness (relative species abundance), percentage cover, percentage frequency and Simpson's diversity index	
• use species diversity indices, species interactions (predation, competition, symbiosis, disease) and abiotic factors (climate, substrate, size/depth of area) to compare ecosystems across spatial and temporal scales	
• explain how environmental factors limit the distribution and abundance of species in an ecosystem.	
• Mandatory practical: Determine species diversity of a group of organisms based on a given index.	
Classification processes	
• recognise that biological classification can be hierarchical and based on different levels of similarity of physical features, methods of reproduction and molecular sequences	

››

• describe the classification systems for – similarity of physical features (the Linnaean system) – methods of reproduction (asexual, sexual – K and r selection) – molecular sequences (molecular phylogeny – also called cladistics)	
• define the term *clade*	
• recall that common assumptions of cladistics include a common ancestry, bifurcation and physical change	
• interpret cladograms to infer the evolutionary relatedness between groups of organisms	
• analyse data from molecular sequences to infer species evolutionary relatedness	
• recognise the need for multiple definitions of species	
• identify one example of an interspecific hybrid that does not produce fertile offspring (e.g. mule, *Equus mulus*)	
• explain the classification of organisms according to the following species interactions: predation, competition, symbiosis and disease	

›› • understand that ecosystems are composed of varied habitats (microhabitat to ecoregion)	
• interpret data to classify and name an ecosystem	
• explain how the process of classifying ecosystems is an important step towards effective ecosystem management (consider old-growth forests, productive soils and coral reefs)	
• describe the process of stratified sampling in terms of: – purpose (estimating population, density, distribution, environmental gradients and profiles, zonation, stratification) – site selection – choice of ecological surveying technique (quadrats, transects) – minimising bias (size and number of samples, random-number generators, counting criteria, calibrating equipment and noting associated precision) – methods of data presentation and analysis.	
• Mandatory practical: Use the process of stratified sampling to collect and analyse primary biotic and abiotic field data to classify an ecosystem.	

Exam practice

Multiple-choice questions

Each multiple-choice question is worth 1 mark.

Solutions start on page 152.

Question 1

A clade is defined as a

A single organism that is a common ancestor.

B single organism that is a lineal descendant.

C group of organisms that consists of a common ancestor and all its lineal descendants.

D group of organisms that consists of a common ancestor and only living descendants.

Question 2

A limitation of the biological species concept is that it does not

A allow for species to have genetic variation.

B account for species that reproduce sexually.

C account for species that reproduce asexually.

D allow for species to have a common ancestor.

Question 3

The distribution represented in the following diagram is best described as

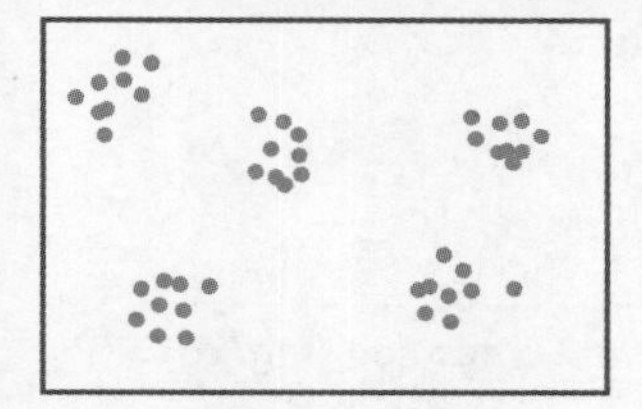

A bare.

B clumped.

C random.

D uniform.

Question 4

Random sampling is used when

A there is an obvious gradient in the sample area.

B species in the sample area are uniformly distributed.

C only information about species distribution is required.

D the sample area is small and there is unlimited time for sampling processes.

Question 5

The ACFOR scale is a simple scale used to describe species abundance within a given area, usually determined using a quadrat. The following table summarises this scale for some species of lichen and algae.

ACFOR scale for lichen and algae in percentages

Species	Abundant	Common	Frequent	Occasional	Rare
Lichen	>80	50–80	20–49	10–19	<10
Alga 1	>90	70–90	40–69	10–39	<10
Alga 2	>90	60–90	30–59	5–29	<5

9780170459136

A researcher found a 60% cover of lichen, alga 1 and alga 2 in the survey area. This means that the abundance of each species is

A abundant, common, frequent.

B common, common, common.

C common, common, frequent.

D common, frequent, common.

Question 6 ©QCAA QCAA 2020 P1 MC Q4

The following information includes:

- a key that is used to classify the types of alpine and subalpine grassland habitats
- a table of abiotic and biotic data obtained from a habitat survey.

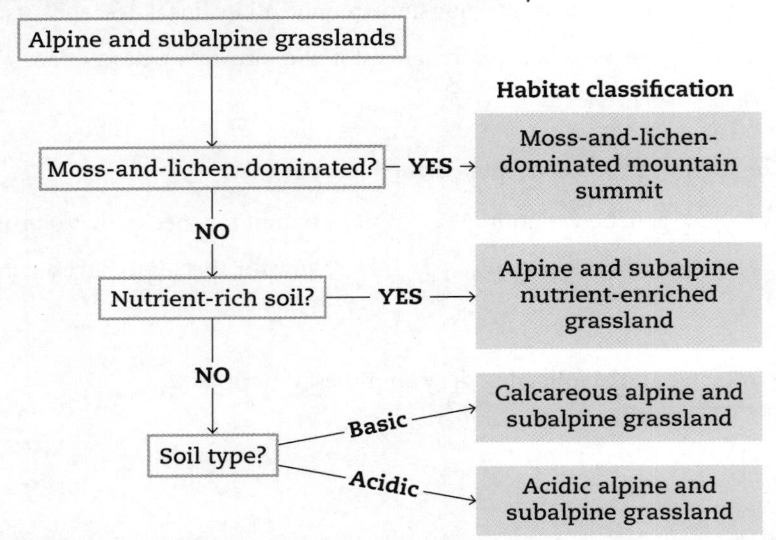

Abiotic physical parameter	Reading	Reference range for nutrient-poor soil (mg/kg)
pH	6.1	
Nitrates/nitrites (mg/kg)	4	<5
Ammonia (mg/kg)	1	<4
Total phosphorus	16	<20

Biotic description: Small amount of low-lying moss, growing on soil substrate. Predominantly low-lying grasses.

Using the data in the table and the key, this alpine and subalpine grassland would be classified as

A a moss-and-lichen-dominated mountain summit.

B an alpine and subalpine nutrient-enriched grassland.

C a calcareous alpine and subalpine grassland.

D an acidic alpine and subalpine grassland.

Question 7 ©QCAA QCAA 2020 P1 MC Q16

The figures show the original distribution zone of Species I and some possible distribution zones of Species I after the introduction of Species II.

Species II has a competitive advantage over Species I; however, it does not tolerate areas of lower rainfall.

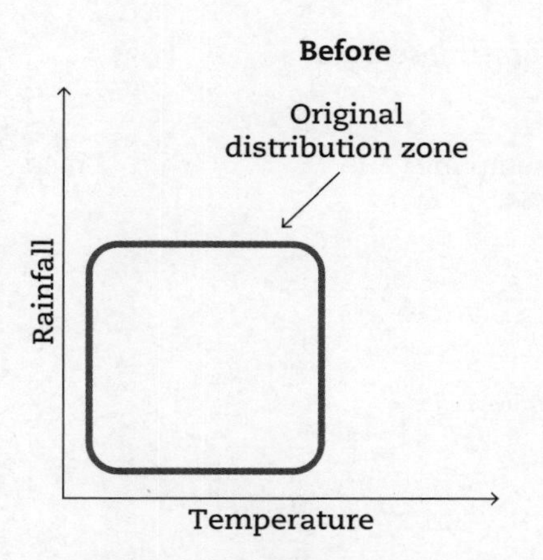

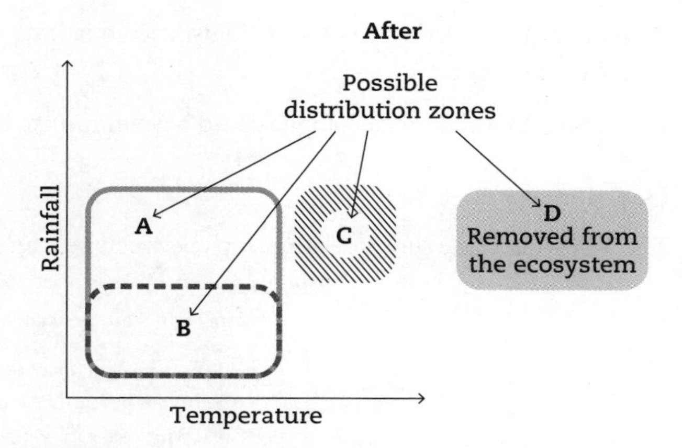

Which of the following would be an accurate prediction of the new distribution zone for Species I?

A Zone A, i.e. no change to the distribution

B Zone B, i.e. reduced distribution within the original zone

C Zone C, i.e. new distribution outside the original zone

D Zone D, i.e. complete removal of the species from the ecosystem

Short-response questions

Solutions start on page 152.

Question 8 (9 marks)

Below is an extract of Specht's classification table.

Life form of tallest stratum	**Projective foliage cover of the tallest stratum**				
	>70%	**51–70%**	**31–50%**	**10–30%**	**<10%**
Trees >30 m	Tall closed forest	Tall forest	Tall open forest	Tall woodland	n/a
Trees 10–30 m	Closed forest	Forest	Open forest	Woodland	Open woodland
Trees <10 m	Low closed forest	Low forest	Low open forest	Low woodland	Low open woodland
Shrubs >2 m	Closed scrub	Scrub	Open scrub	Tall shrubland	Tall open shrubland
Shrubs (S) 0.25–2 m	Closed heathland	Heathland	Open heathland	Shrubland	Open shrubland
Shrubs (NS) 0.25–2 m	n/a	n/a	Low shrubland	Low shrubland	Low open shrubland
Shrubs (S) <0.25 m	n/a	n/a	n/a	Dwarf open heathland	Dwarf open heathland
Shrubs (NS) <0.25 m	n/a	n/a	n/a	Dwarf shrubland	Dwarf open shrubland

An ecologist sampled an area of trees on the Sunshine Coast, using the process of stratified sampling and quadrats.

a One set of the ecologist's data found that the tallest stratum of trees was on average 9 metres. The dominant species was *Melaleuca* (tea-tree). The foliage cover was 20–25%. Classify and name this ecosystem using the information in the table and the ecologist's data. 1 mark

b The ecologist classified a second area as eucalypt open forest. Describe this ecosystem, using the information in the table. 3 marks

c Provide one reason why the ecologist used the process of stratified sampling. 1 mark

9780170459136

d Identify two reasons why the ecologist used quadrats rather than line transects to collect this data. 2 marks

e Identify two ways the ecologist could have minimised bias when sampling. 2 marks

Question 9 (3 marks)

The following clade shows relationships between pelican species.

Pelecanidae

American white pelican, *Pelecanus erythrorhynchos*
Brown pelican, *Pelecanus occidentalis*
Peruvian pelican, *Pelecanus thagus*
Great white pelican, *Pelecanus onocrotalus*
Australian pelican, *Pelecanus conspicillatus*
Pink-backed pelican, *Pelecanus rufescens*
Dalmatian pelican, *Pelecanus crispus*
Spot-billed pelican, *Pelecanus philippensis*

a Circle an example of a clade that includes the Peruvian pelican in the diagram. 1 mark

b Mark the node of the most recent common ancestor of the Australian pelican and the American pelican. 1 mark

c Which pelican is more closely related to the Peruvian pelican – the American white pelican or the Australian pelican? 1 mark

Question 10 (7 marks)

An ecologist sampled a mangrove environment in two different areas to determine the diversity of species present. The data is summarised in the following table.

Summary of species observed at the mangrove sample site

Species	Site 1 numbers	Site 2 numbers
Avicennia marina (grey mangrove)	7	10
Algae	15	5
Fish	20	2
Pelican	6	0

Hint
You will be provided with the rule, but you need to know what the letters represent.

$$\mathrm{SDI} = 1 - \frac{\sum n(n-1)}{N(N-1)}$$

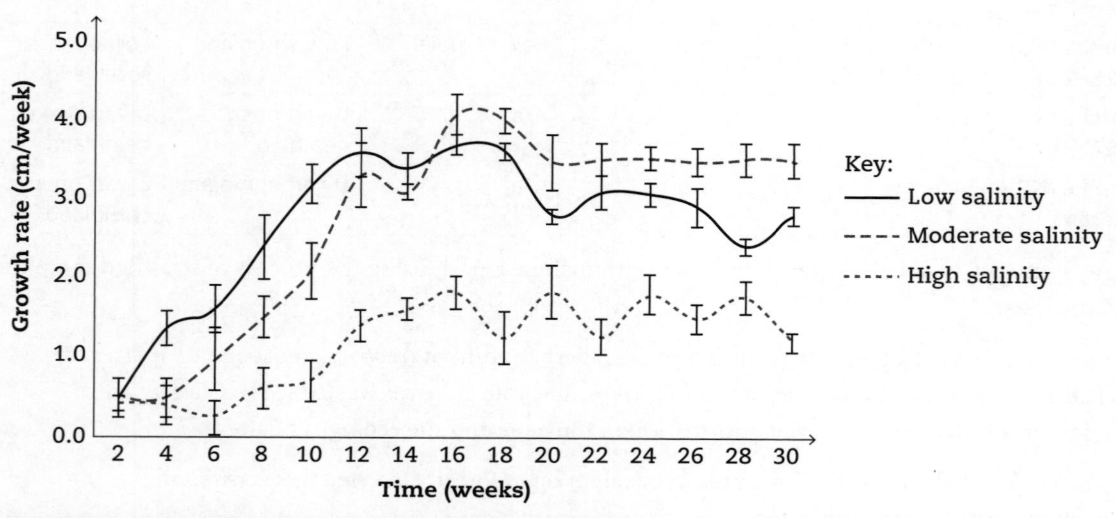

GRAPH 1: The growth rate of *Avicennia marina* at different salinity levels

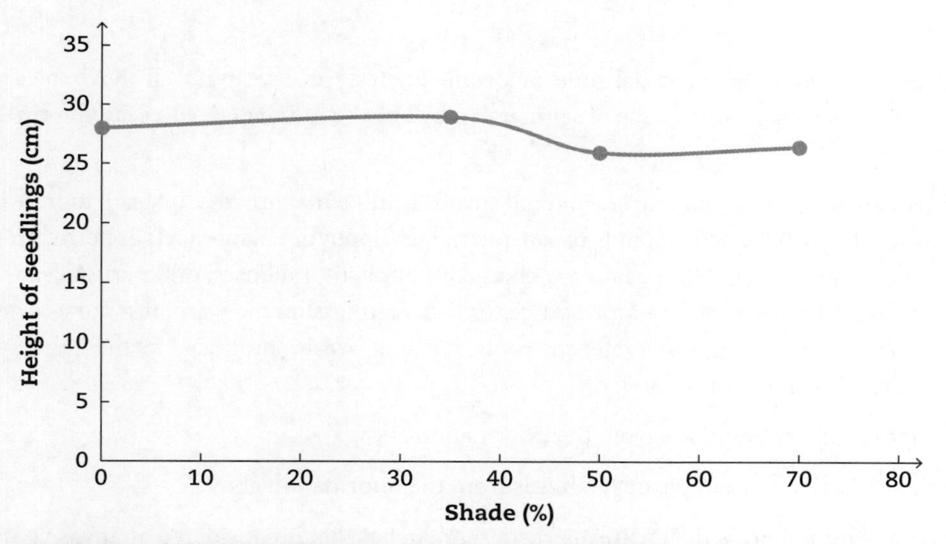

GRAPH 2: The effect of reducing light intensity (shading) on seedling growth of mangroves

a Use the SDI formula to calculate the SDI for site 1. Round your answer to two decimal places. 2 marks

b Compare the species richness of sites 1 and 2. 3 marks

c Explain what factor may limit the abundance of *Avicennia marina* at these sampling sites, using information from the two graphs. 2 marks

Question 11 (6 marks)

The graph summarises the projected life expectancy and percentage of offspring that survive at each life span.

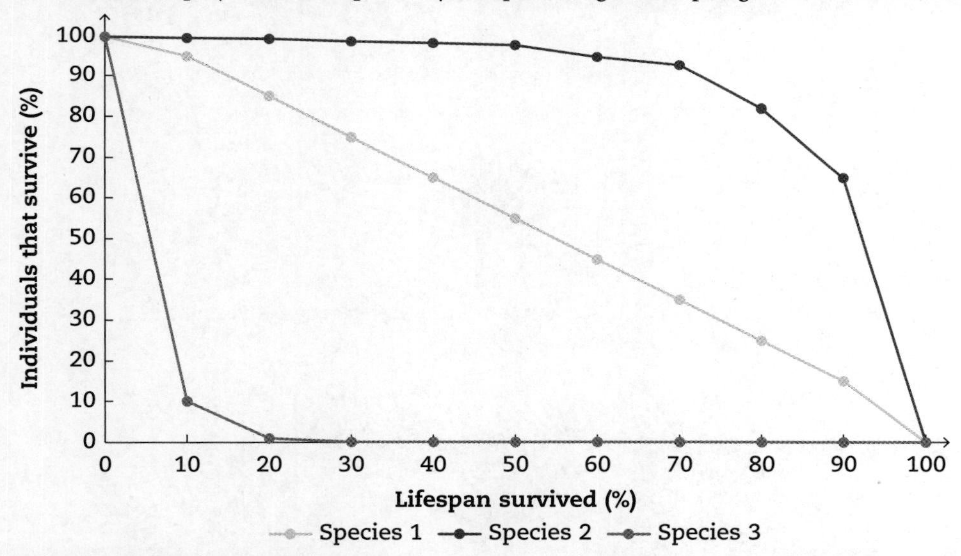

The life expectancy of three different species

An average female koala has a 12-year life span and can breed from 4 years of age. After conception, it is 35 days until the birth of the joey, which then remains in the mother's pouch for 6 months. The joey leaves its mother's care between 1 and 2 years of age. Each female can produce five or six offspring over her lifetime.

a Describe which species represented in the graph is mostly likely to have an asexual method of reproduction. 2 marks

b Which species in the graph is most likely to represent a species of koala? Provide two reasons for your answer. 3 marks

c Classify the koala as an r-selected or K-selected species. 1 mark

Question 12 (5 marks)

Saltmarshes are situated in the intertidal zone, below the highest tide level but well above the low tide level along the Queensland coast. Some areas of saltmarsh may only be inundated with seawater by the highest spring tides.

Saltmarshes are usually devoid of trees and tall shrubs (unlike mangroves). Instead, to trap and stabilise sediments they have dense stands of salt-tolerant (halophytic) plants such as herbs, grasses and low shrubs. They support invertebrate fauna such as crabs, prawns, molluscs, other crustaceans and worms. These invertebrates provide a food source for fish. Saltmarshes may also be a nursery area for fish. Crabs dig to make burrows and this aerates the roots of plants. Wastes produced by these invertebrates increase the nitrogen content of the soil.

a Identify an example of predation from the information above. 1 mark

b Describe and classify an example of symbiosis from the information above. 2 marks

c *Spartina* is a common grass in North American saltmarshes that is considered an invasive species in Australia. Its presence transforms saltmarshes into '*Spartina* meadows', altering the availability of food species, reducing numbers of wading birds and decreasing the diversity of plants.

Explain whether the introduction of *Spartina* is considered interspecific or intraspecific competition. 2 marks

Question 13 (6 marks)

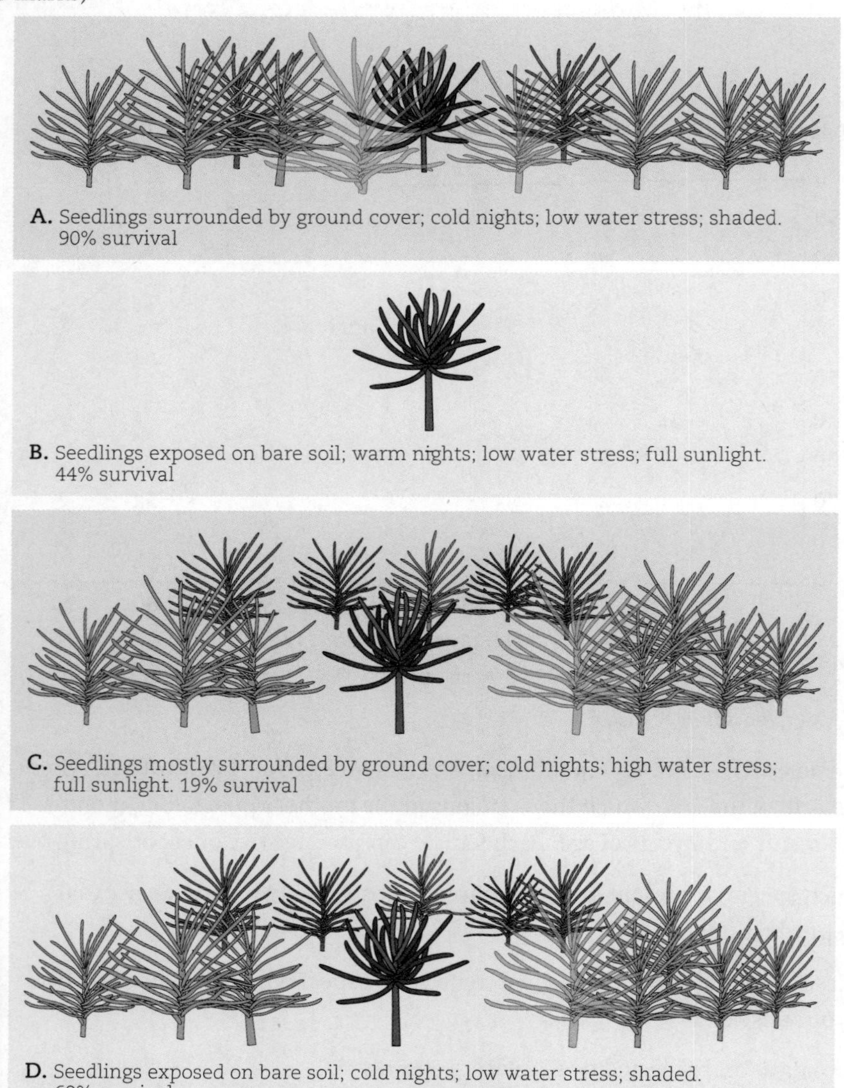

Data from microhabitat experiment on survival of pine tree seedlings in alpine Australia

A botanist was researching the best conditions to germinate pine tree seedlings native to Australia. The seedlings were under the conditions described in the diagram. The percentage survival of seedlings was recorded.

Explain which environmental factor(s) had the greatest effect on seedling survival.

Question 14 (4 marks) ©QCAA QCAA 2020 P1 SR Q22

The following cladogram proposes the evolutionary history of several fish phyla (A–H).

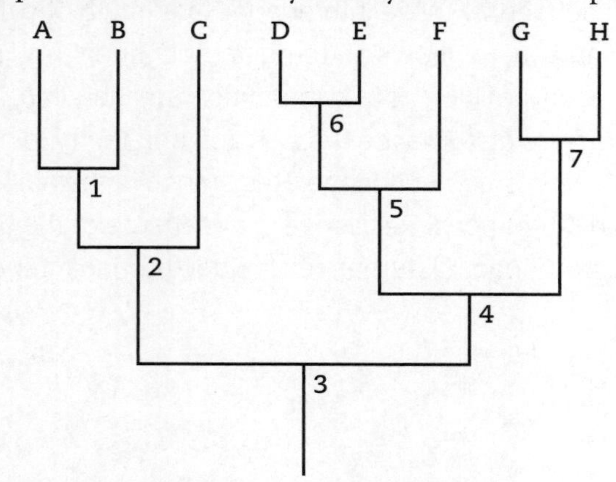

a Define the term *clade*. Circle an example of a clade on the cladogram. 2 marks

b Which node (1–7) represents the most recent common ancestor for species E and G? 1 mark

c Which two species shared the most recent common ancestor? 1 mark

Question 15 (4 marks) ©QCAA QCAA 2020 P2 SR Q1

a Explain how a species interaction may be classified as symbiotic. 2 marks

b Using an example, describe the symbiotic relationship of mutualism. 2 marks

Chapter 2
Topic 2 Ecosystem dynamics

Topic summary

An ecosystem is a community of living organisms interacting with the non-living components of the environment. Ecosystems are defined by these interactions, and the biotic and abiotic components are linked through nutrient cycles and energy flows. External factors such as climate and soil structure, and internal factors such as decomposition, root competition and shading, affect the availability of resources. Other internal factors are disturbance, succession and the types of species present.

In a system at equilibrium, there is no net change. Resistance is the ability of the ecosystem to remain at equilibrium despite disturbances. Resilience is a measure of the time it takes for a system to return to equilibrium after a disturbance. Disturbances include human interventions and catastrophic weather events.

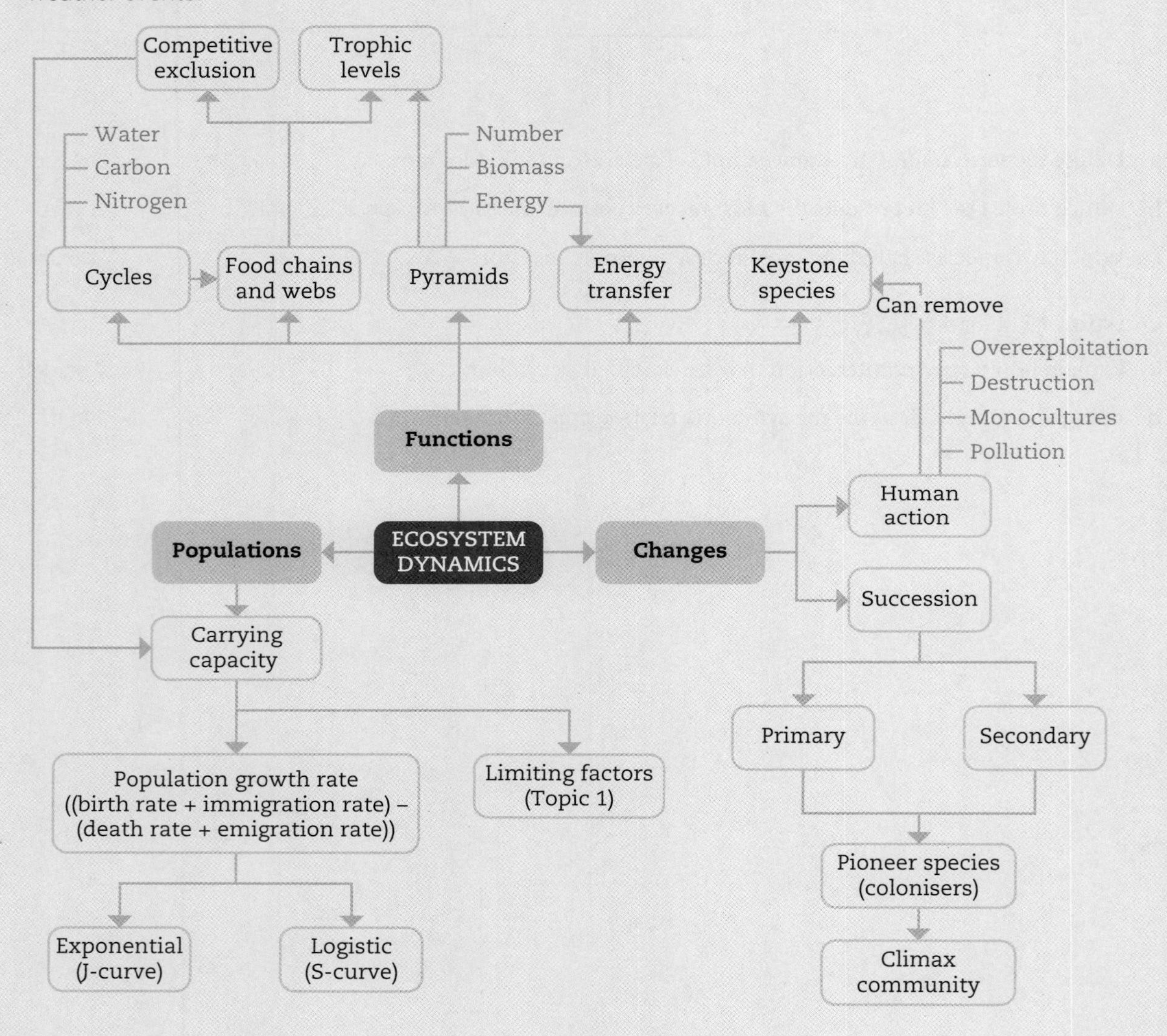

2.1 Functioning ecosystems

2.1.1 Energy transfer and transformation

The source of energy for most ecosystems is the Sun. **Photosynthesis** is the process that transforms light energy into chemical energy (glucose). Light energy is absorbed by chlorophyll in the cells of photosynthetic organisms, e.g. plants and algae.

$$6CO_2 + 6H_2O \xrightarrow{\text{light energy}} C_6H_{12}O_6 + 6O_2$$

where CO_2 is carbon dioxide, H_2O is water, $C_6H_{12}O_6$ is glucose, O_2 is oxygen, and the reaction requires light energy.

The chemical energy is then **transferred** and **transformed** through the ecosystem. These energy movements and changes are summarised by food webs, food chains and ecological pyramids. Energy, measured in kilojoules (kJ), is not recycled and needs to be acquired continuously.

Energy moves through **trophic levels** (TL), or feeding levels, from autotrophs to heterotrophs then finally to **decomposers**. **Autotrophs** are also called primary **producers**. **Heterotrophs** are primary, secondary or tertiary **consumers** depending on whether they are herbivores, carnivores or omnivores. Each TL receives around 10% of the available energy from the previous TL. Of the available energy, some is stored as fat or protein, most is converted into heat energy and only about 10% of the available energy moves from one trophic level to the next. Heat energy cannot be converted into other forms by these organisms.

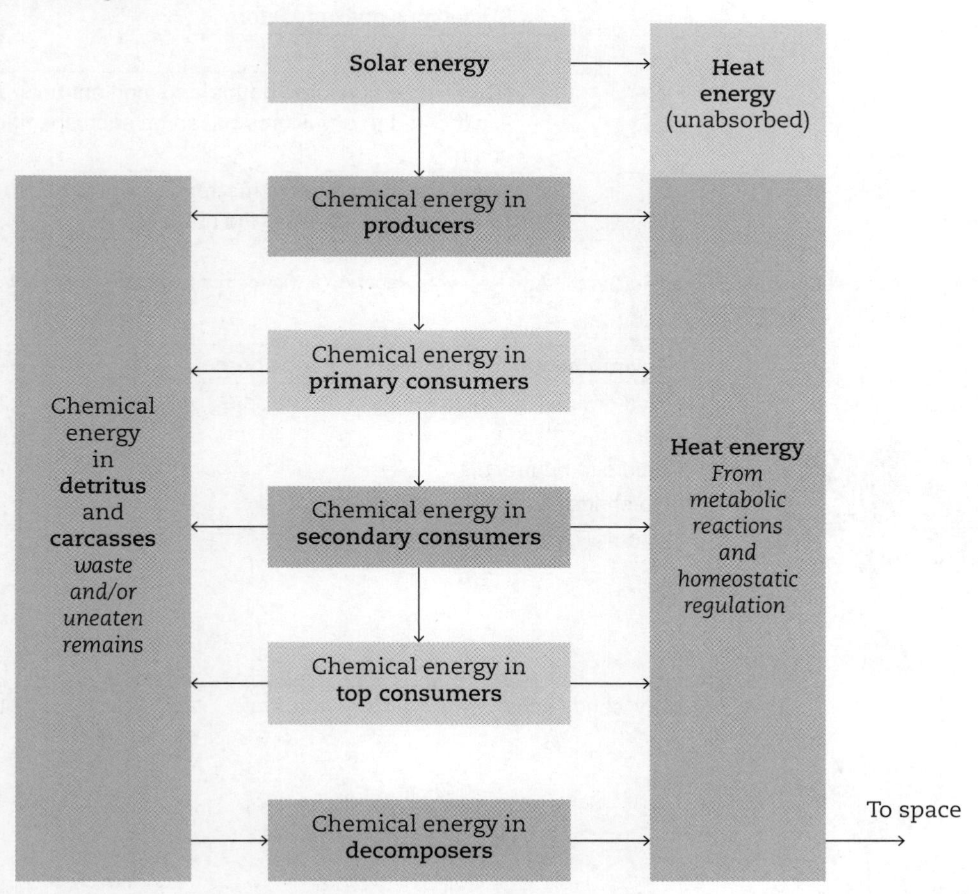

FIGURE 2.1 Transfer and transformation of energy in an ecosystem

All energy converted into biomass by primary producers is called **gross primary productivity** (GPP). Some of this energy is used for respiration. The remaining energy is called **net primary productivity** (NPP) and is available to be consumed by heterotrophs. Net productivity is the energy available to TL2 and above. Primary productivity is measured in grams of carbon per square metre per year or kJ m^{-2} $year^{-1}$.

9780170459136

Low GPP	**High GPP**
Lower temperatures, lower light intensity, nutrient poor	Higher temperatures, higher light intensity, nutrient rich
Desert	Tropical rainforest Coral reef

FIGURE 2.2 Primary productivity in different ecosystems

A trophic level consists of organisms that carry out the same role in the **food web**, acting as the source of nutrition for the level above.

TABLE 2.1 Summary of trophic levels in an ecosystem

Organisms	Typical trophic level (TL)	Description
Producers – autotrophs	1	Autotrophs that produce their own food through photosynthesis (Some single-celled organisms are chemoautotrophs that produce their own food from inorganic substances)
Consumers – heterotrophs	2	Herbivores that consume plant material only (primary consumers)
	3	Carnivores that consume herbivores (secondary consumers)
	4	Carnivores that consume other carnivores (tertiary consumers) Sometimes apex predators
	5	Apex predators
	2+*	*Omnivores consume both plants and animals. The TL is allocated to omnivores based on each specific food chain.
Decomposers and **detritovores**		Consume dead organic matter and wastes from all trophic levels, recycling the nutrients

TABLE 2.2 A data table summarising organisms in the Arid Recovery Reserve (a wildlife reserve in the north of South Australia) as an example of an ecosystem

Species	Environmental requirements	Food source	TL
Acacia aneura (mulga)	Well-drained sandy to loamy soils Acidic to neutral pH No shade About 370 mm rain per year	Autotroph	1
Sedge	Dry, sandy soils	Autotroph	1

››

Species	Environmental requirements	Food source	TL
Ant	Limited data available on Australian ants	*Acacia* seeds	2
Rufous whistler	Mallee and scrub of arid interior, less common in wetter tall forests	Insects, seeds	3
Bilby	Prefers spinifex and acacia habitats Nocturnal Burrows to 2 m to keep cool during day Low rainfall, high daily temperatures (30–40°C)	Insects and their larvae, seeds, spiders, bulbs, fruit, fungi	2
Shingleback lizard	Ideal temperature 30–35°C Forests, shrublands and desert grasslands to sandy dunes	Berries, fruit, flowers	2 and 3
Kookaburra	Open woodland and forests	Insects, reptiles, amphibians, and small mammals, snakes up to 1 m	3 and 4
Dingo	Deserts where waterholes are available Wide range of temperatures	Apex predator – eats small mammals	3
Rabbit	Requires sandy soils to dig warren Prefers short grasses, but can live in many different environments	Plant matter with less than 40% fibre with 10–12% protein for maintenance and 14% protein for reproduction Obtain water from green vegetation	2

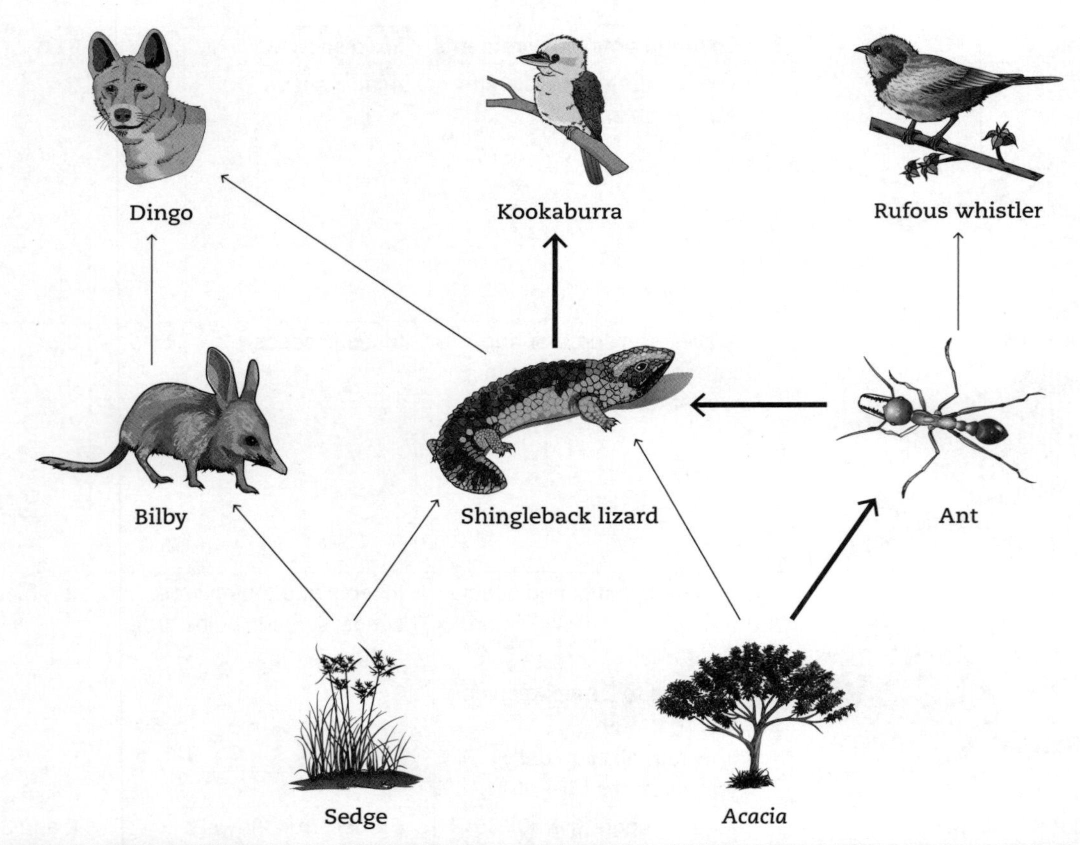

FIGURE 2.3 This food web, based on information in Table 2.2, shows the transfer of energy and matter. Notice that some organisms can occupy more than one trophic level. The series of bold arrows represents one **food chain** within this food web.

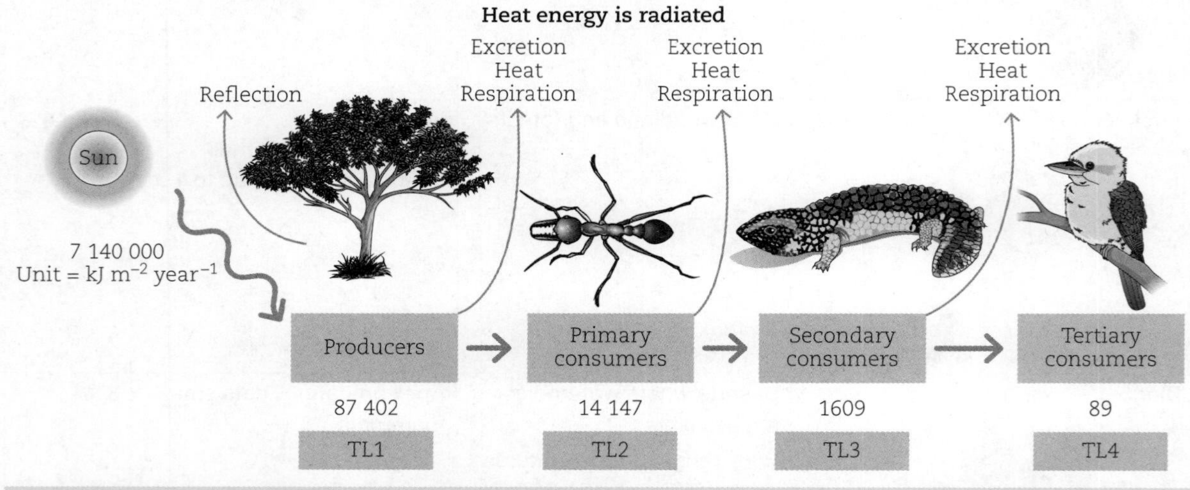

FIGURE 2.4 The transfer of energy between trophic levels after producers have converted light energy into chemical energy for a single food chain. Note that the energy transfer is not efficient.

Hint

Note that different sources of data will use different units to 'measure' the transfer of energy or biomass, e.g. net productivity or energy available.

Plants typically capture about 1.5% of the Sun's energy, converting about 1% of the energy into biomass available to TL2.

The **energy efficiency** for each transfer can be calculated by the equation:

$$\text{efficiency} = \frac{\text{energy transferred to next level}}{\text{total energy in}} \times 100$$

In Figure 2.4, the energy available to TL2 primary consumers is 14 147 kJ m^{-2} $year^{-1}$ and in TL3 it is 1609 kJ m^{-2} $year^{-1}$.

$$\text{efficiency} = \frac{1609}{14\,147} \times 100 = 11.4\%$$

Therefore, in this system TL3 receives 11.4% of the energy available in TL2.

On average, about 10% of the energy at one TL is available to the next TL. The amount of energy transferred decreases the more TLs there are. Energy is lost through respiration, excretion and as heat energy. These are the result of the processes that keep organisms alive. Figure 2.5 is a more detailed diagram that includes the energy lost to these processes.

Another way of representing the movement of energy in an ecosystem is to use a food chain that represents the number of individuals, biomass or energy at each trophic level.

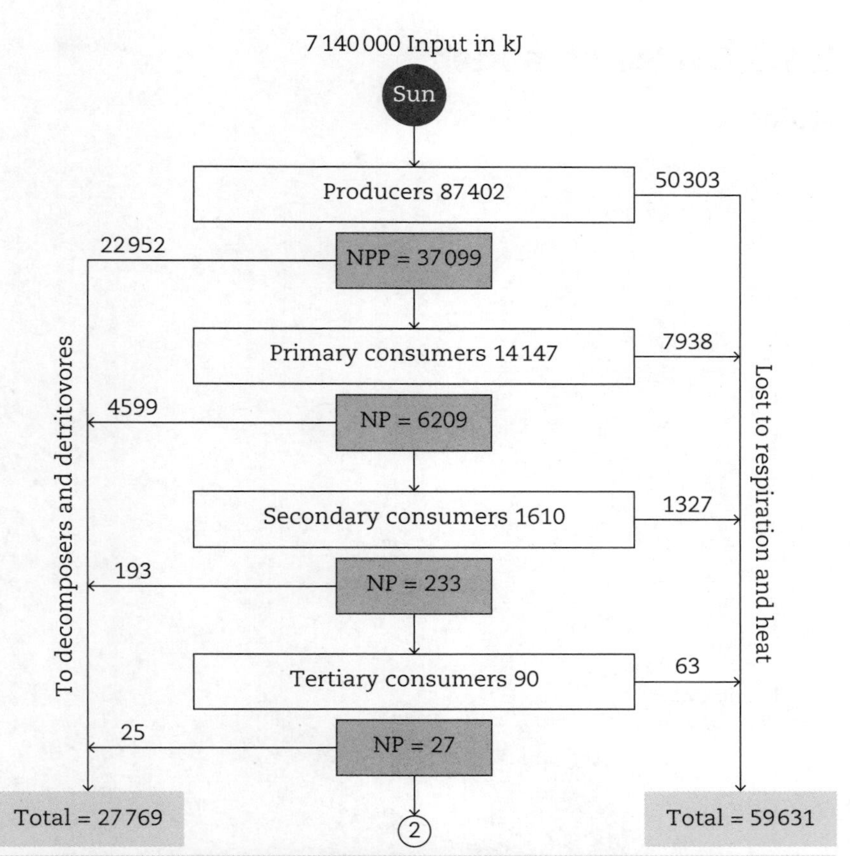

FIGURE 2.5 An example of energy transfer in an ecosystem with four trophic levels. Note that by TL4 there is very little energy available for the consumers. NPP, net primary productivity; NP, net productivity.

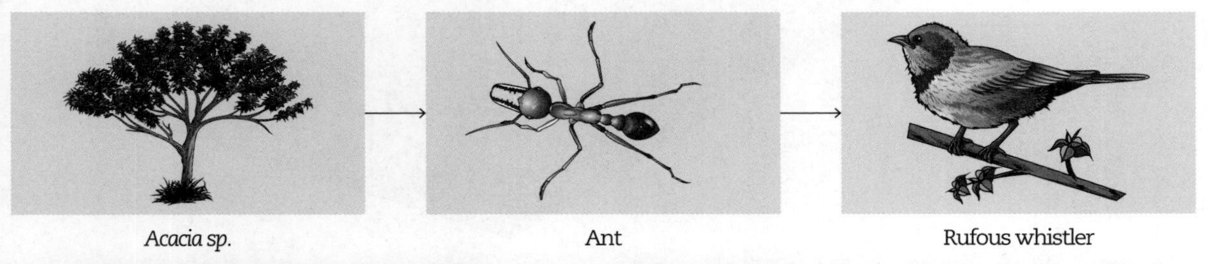

FIGURE 2.6 A simplified food chain showing the transfer of energy and matter through three trophic levels. The *Acacia* is a producer at TL1, the ant is a primary consumer (herbivore) at TL2, and the rufous whistler is a secondary consumer at TL3.

This food chain can be represented as an **ecological pyramid** (Figure 2.7).

Pyramids of number may be upright or inverted, as shown in Figure 2.7.

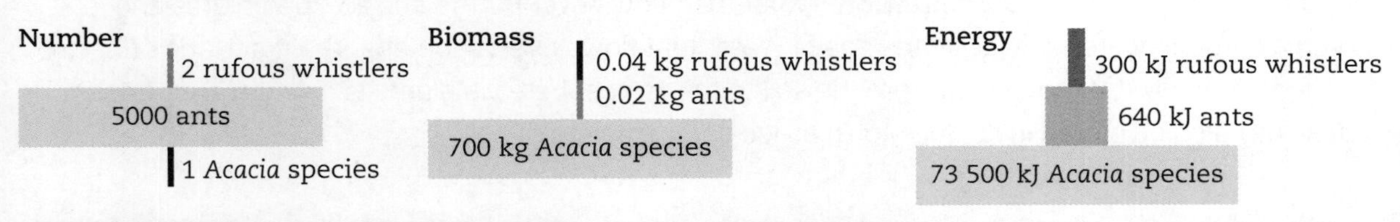

FIGURE 2.7 Three ecological pyramids all based on the same food chain. Only the energy pyramid has the 'typical' shape. Note that these pyramids are not to scale.

Hint
We could also represent the entire food web as a pyramid.

Hint
Think field of grass versus a tree.

Pyramids of biomass for an ecosystem are usually upright. Ecosystems on plankton can have a low producer biomass due to a high reproduction rate. Pyramids of energy are always upright. Without sufficient GPP, the system cannot be supported. However, pyramids of energy account for productivity. Note that the energy efficiency may not be 10% for a food chain because it is taking into account only one species at each TL.

2.1.2 Cycles of matter

Water

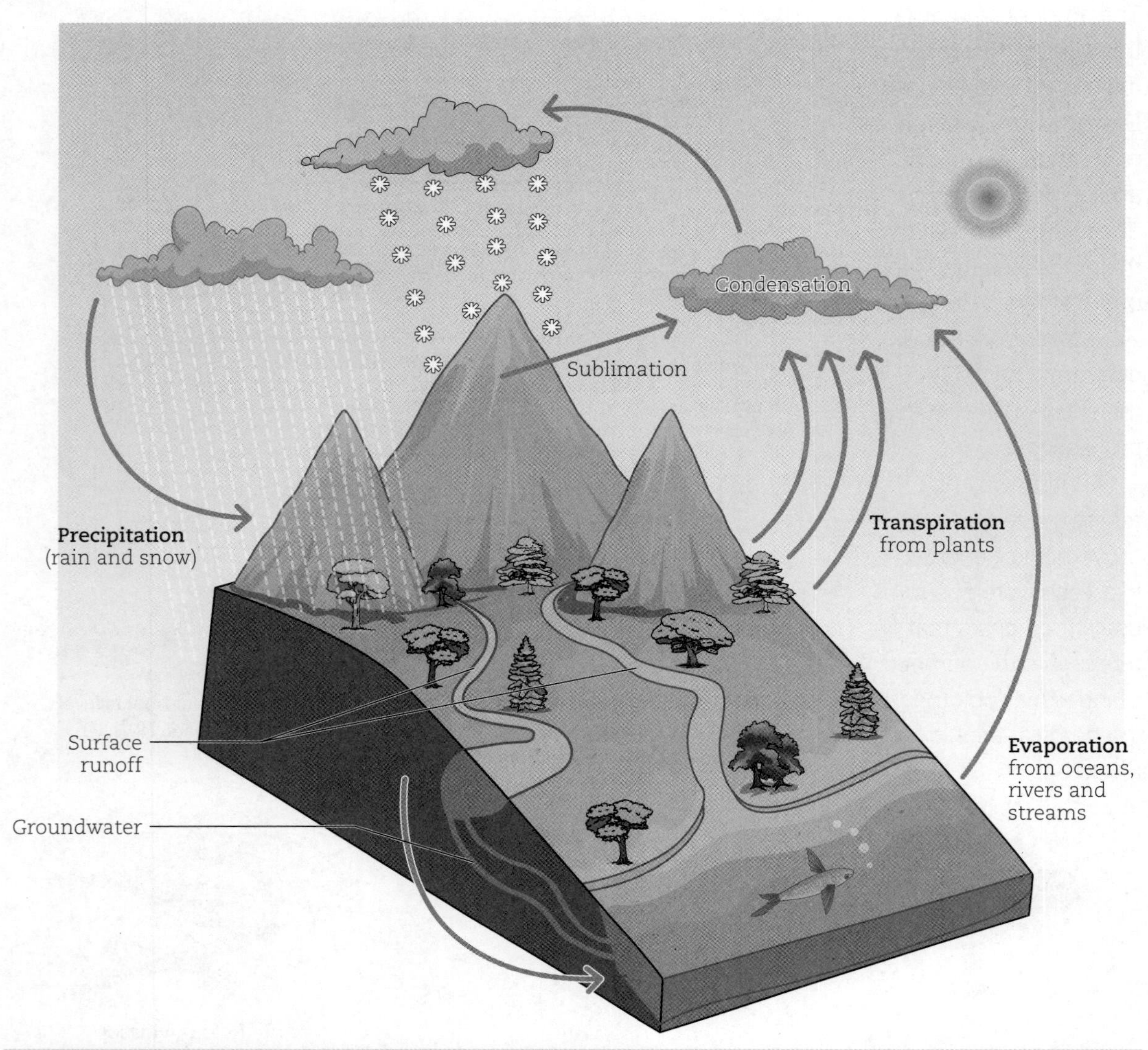

FIGURE 2.8 The water cycle, showing the processes involved in moving water through ecosystems. The water cycle is the continual circulation of water from clouds to the land, to the ocean, and back to the clouds.

The Sun is the primary source of energy for **evaporation** and sublimation to move water from the ocean to the atmosphere as water vapour. The vapour cools and condenses. **Condensation** of water forms clouds that result in **precipitation**. Water then flows over Earth's surface, percolates and is stored as ground water or evaporates again. Water that flows over a non-absorbent surface of Earth is called runoff. Water movement from the soil to the atmosphere via plants is called **transpiration**. Water may also be stored in its solid form as ice caps, glaciers or snow.

Carbon

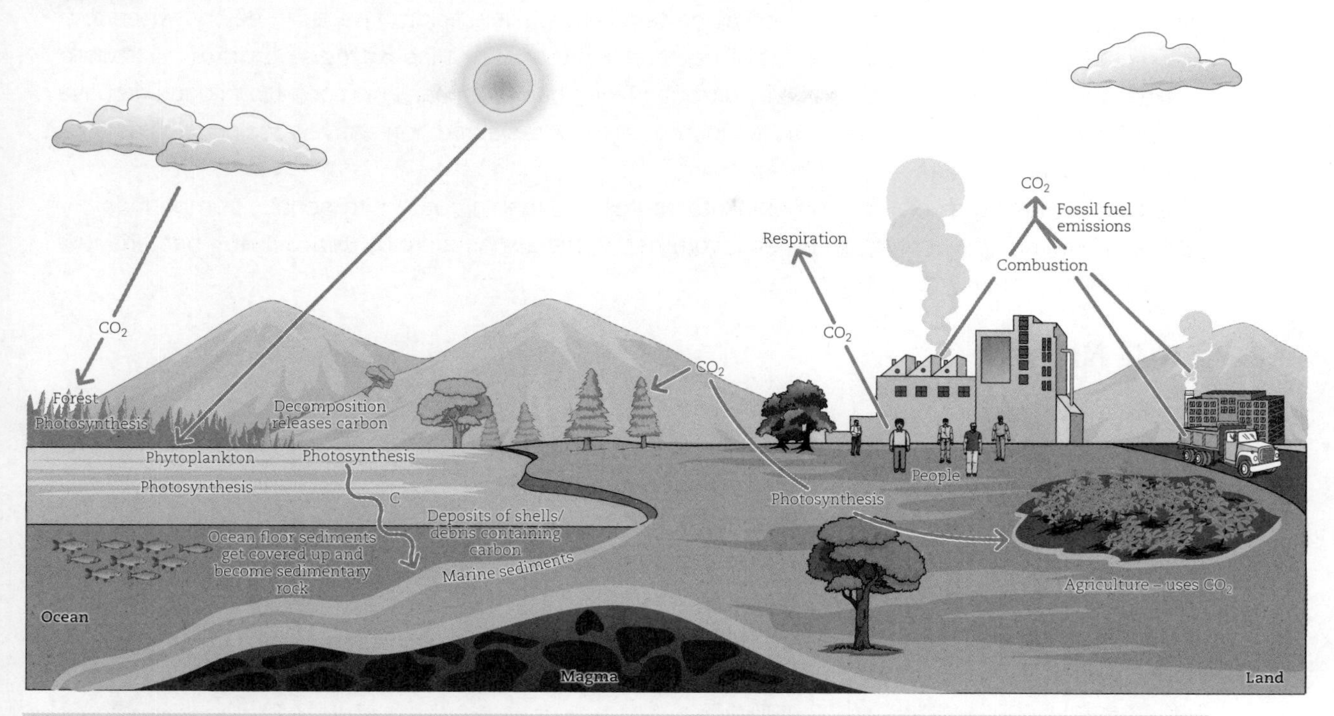

FIGURE 2.9 The carbon cycle, showing the processes involved in moving carbon through ecosystems

Carbon atoms are cycled between Earth and the atmosphere through three main processes: photosynthesis, **respiration** and **combustion**. Rocks, sediments and oceans store most carbon, and are known as carbon **sinks**. Photosynthesis removes carbon, in the form of carbon dioxide, from the atmosphere and converts it into glucose. Carbon is used by living things for cellular growth. Ideally, the amount of carbon released from one reservoir (**source**) is the same as the amount absorbed by a different reservoir (sink). Respiration and combustion both release carbon from sources.

Nitrogen

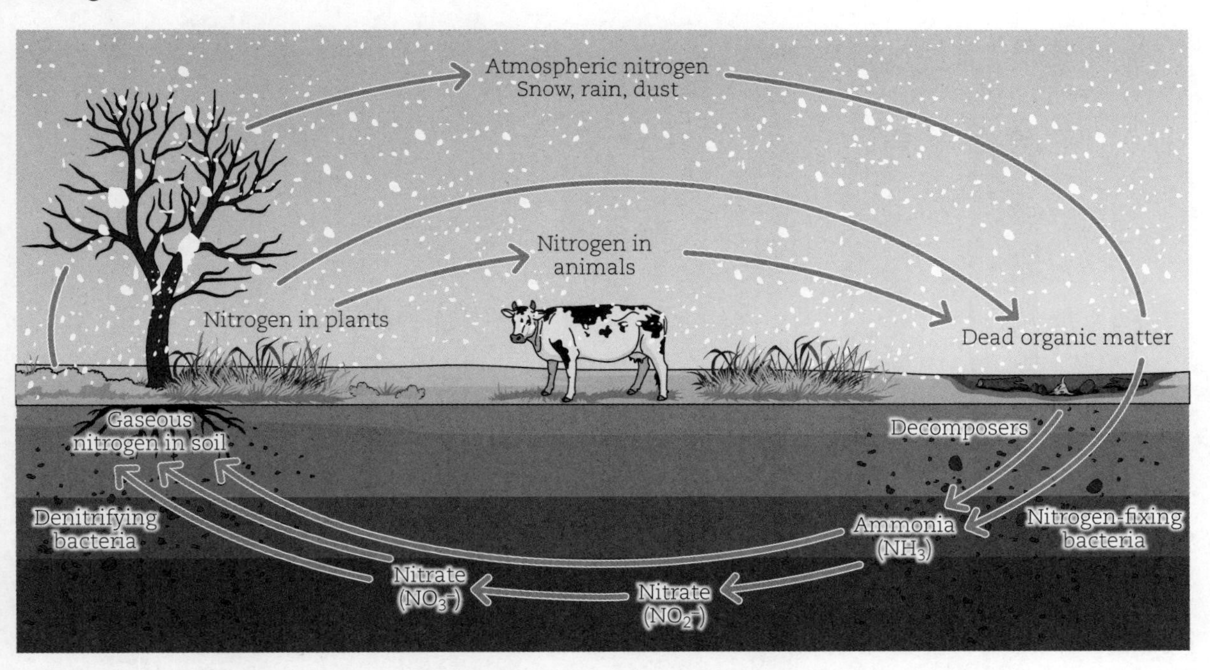

FIGURE 2.10 The nitrogen cycle, showing the processes involved in moving nitrogen through ecosystems

Nitrogen as N_2 gas makes up about 78% of the atmosphere. So that most organisms can access this nitrogen to produce molecules such as proteins and nucleic acids, it needs to be converted to biologically accessible compounds by soil bacteria in a process called **nitrogen fixation**. Nitrogen is 'fixed' and converted into ammonia by nitrogen-fixing bacteria. Many nitrogen-fixing bacteria have symbiotic relationships with plants called legumes (e.g. *Acacia* and non-native species such as clover). Nitrogen can also be fixed by lightning.

Nitrifying bacteria convert ammonia into nitrites and then nitrates, nitrogenous compounds that can be taken up by plants. Nitrogen is returned to the atmosphere by **denitrifying** bacteria that convert nitrates back into nitrogen gas.

2.1.3 Niches

A **niche** is the specific range of environmental roles, conditions and biological interactions required for an organism to survive and reproduce. Niche dimensions can be visualised in a graph such as Figure 2.11, where each axis represents an environmental condition or resource to give the fundamental niche (the total possible range of the species). Species can interact when their fundamental niches overlap. When biological interactions are also considered, we can identify the organism's realised niche (the actual range of the species).

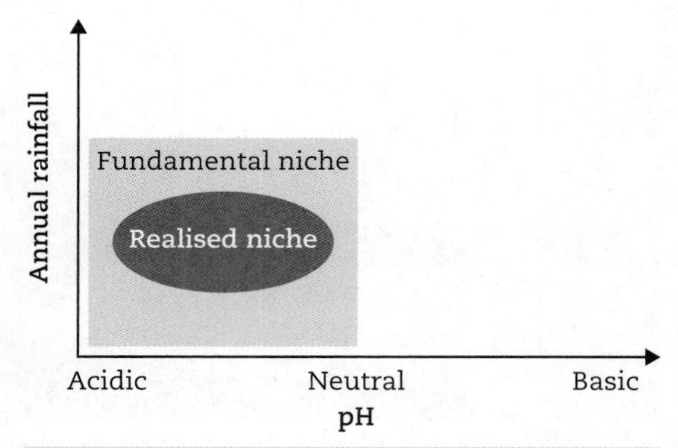

FIGURE 2.11 An example of the difference between the fundamental and realised niche of an *Acacia* species. Note that there are more than two conditions affecting an organism's niche, but only two axes on the graph.

> **Hint**
> Remember competition from Unit 3 Topic 1 – this type of competition is interspecific.

The **competitive exclusion principle** states that species occupying the same niche cannot occupy the same environment for a long time. They are competing for resources. The competitor that has an advantage is more likely to survive.

A common example of competitive exclusion is the growth of two species of *Paramecium*, a single-celled microorganism that lives in aquatic environments. *Paramecium* species prefer warm freshwater pools without currents.

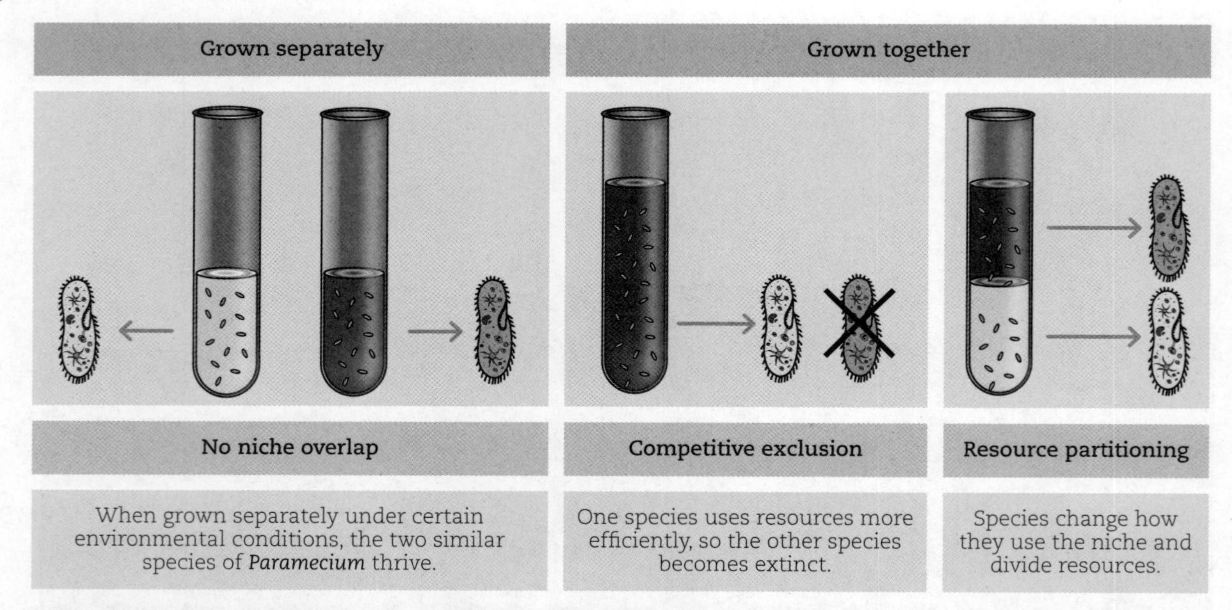

FIGURE 2.12 Possible outcomes when two species of *Paramecium* with similar niches are grown together

Refer to Table 2.2 (page 46) showing the organisms in the Arid Recovery Reserve in South Australia. Notice that the bilby and the rabbit both dig burrows or warrens. Both require plant matter for food. In these two dimensions, rabbits and bilbies have a significant overlap.

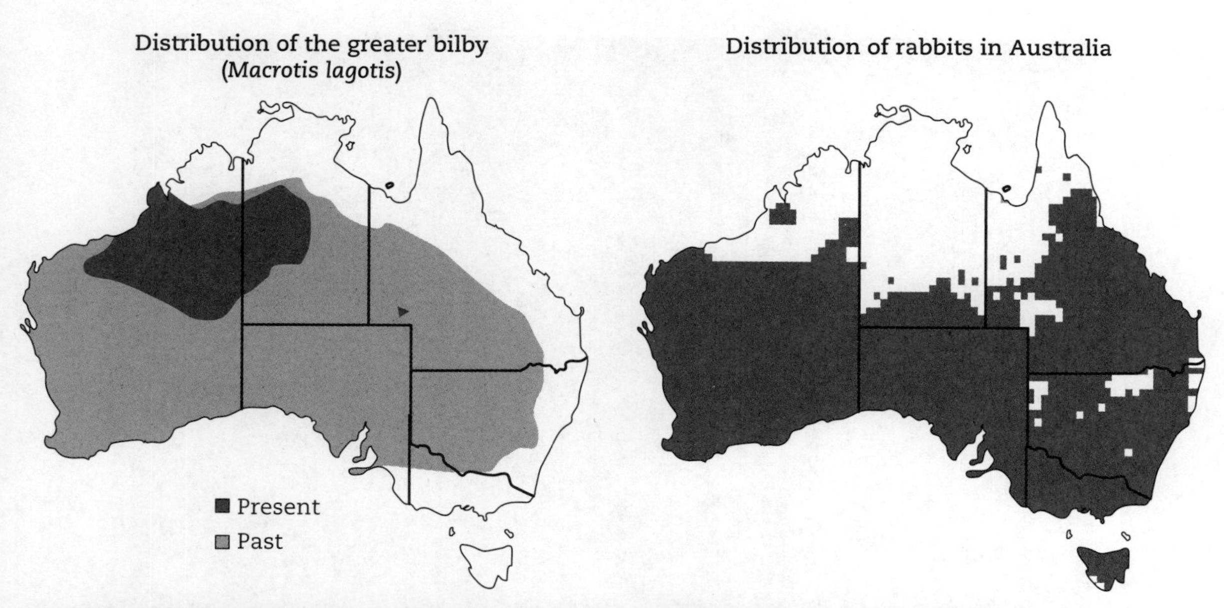

FIGURE 2.13 Overlap of bilby and rabbit distributions, with the subsequent reduction in bilby distribution

2.1.4 Keystone species

A **keystone species** is a plant or an animal that has a unique and crucial role in the way an ecosystem functions. They may affect abundance or type of species in the habitat. They could be a predator, prey, an ecosystem engineer, a mutualist or a plant. The removal of a keystone species results in changes to the habitat structure and biodiversity. Australian keystone endangered species include the cassowary, Gilbert's potoroo, grey nurse shark, Tasmanian devil, red-tailed black cockatoo, Tasmanian wedge-tailed eagle, great white shark, northern quoll and Australian sealion.

In the Arid Recovery Reserve food web (Figure 2.3, page 48), *Acacia* is a keystone species because it can capture, retain and cycle water, nutrients and sediments. *Acacia* pollen and sweet secretions are a food source for many insects and birds. Without this plant species, the structure and diversity of this habitat would change significantly because no other plant performs this role in this system.

Hint
What role do you think each of the keystone species mentioned has in their habitat?

2.2 Population ecology

2.2.1 Population growth and change

Population ecology is the study of how and why populations change over time. Ecologists look at the growth, development, reproduction and survival of individuals to explain species numbers and distributions. 'Population' can be applied on a small or a large scale; for example, ants on a particular *Acacia* tree versus all the ants in the Arid Recovery Reserve of South Australia.

Populations can be described by their:

- size (total number of individuals)
- density (number of individuals per unit area)
- growth (how the number of individuals changes over time).

Ecologists are rarely able to count entire populations. However, populations of mobile organisms can be sampled by a capture–mark–recapture method. The data collected can be used to estimate population size (N), using the Lincoln index:

$$N = \frac{M \times n}{m}$$

where M is the number of individuals caught, marked and released initially; n is the number of individuals caught on the second sampling; and m is the number of recaptured individuals that were marked.

Hint
Remember the reasons we sample from Unit 3 Topic 1.

Hint
Capital letters apply to the bigger numbers of the first sample/population estimate. The lower-case letters apply to the smaller numbers with the second sample. '*M*' or '*m*' always applies to marked individuals.

FIGURE 2.14 A bettong

Worked example

An ecologist was interested in the number of bettongs surviving in the Arid Recovery Reserve. In the first sample, 23 bettongs were caught, tagged and released. In the second sample, 16 bettongs were caught, 6 of which were tagged. $N = ?$ $M = 23$ $n = 16$ $m = 6$ $N = \frac{23 \times 16}{6} = 61.3 = 61$ bettongs	Limitations and assumptions: • That the marking process does not affect the survival of the organism. • There is no immigration or emigration between samplings. • There are no births or deaths between samplings. • That marked individuals can disperse between samplings.

Populations can increase in number through births and immigration. They decrease in number through deaths and emigration. One way of calculating population growth rate and change is:

Population growth rate = (birth rate + immigration rate) – (death rate + emigration rate)

Worked example

> Nine bilbies were released into an area in an attempt to increase their population. Over a two-year period, 22 bilbies were born and survived. Two bilbies were killed by feral cats. No bilbies moved in or out of the area. This means:
>
> Population growth rate = (birth rate + immigration rate) – (death rate + emigration rate)
>
> Birth rate = 22 bilbies ÷ 2 years = 11 bilbies per year
> Death rate = 2 bilbies ÷ 2 years = 1 bilby per year
> Immigration rate = 0 bilbies ÷ 2 years = 0 bilbies per year
> Emigration rate = 0 bilbies ÷ 2 years = 0 bilbies per year
> Population growth rate = (11 + 0) – (1 + 0) = 10 bilbies per year.
>
> *Extension:*
>
> The original population was 9 bilbies. With a growth rate of 10 bilbies per year, after two years the population should be 29 bilbies.

Hint

A rate is a value per unit of time. If 20 individuals are born in a year, then birth rate is 20 births/year.

Each organism has a set of conditions that allow for a maximum population growth rate. In this instance the growth is considered **exponential**. This is called a J-curve. However, populations generally cannot maintain a high growth rate. Eventually, population growth is limited by competition for resources, disease, food availability or natural disasters. If the growth rate is reduced by a limiting factor and slows to zero, it results in an S-shaped or **logistic** growth curve (Figure 2.15).

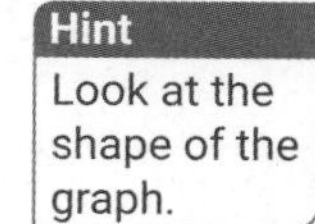

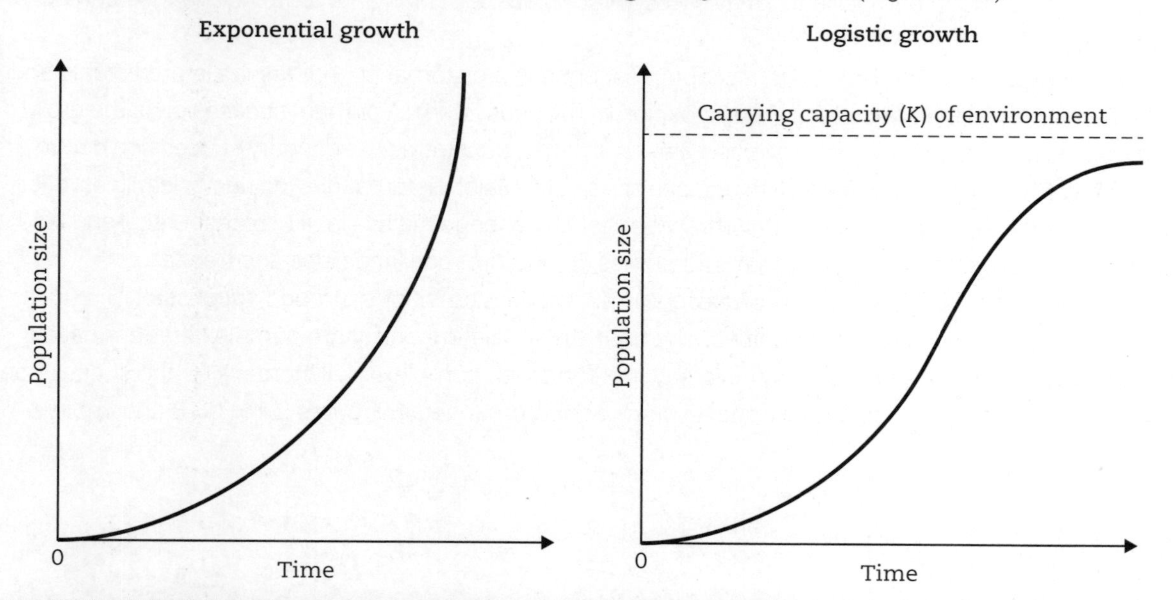

FIGURE 2.15 Graphs comparing the outcomes of exponential growth with logistic growth

Two types of factors affect population growth – **density-dependent** and **density-independent** factors (Table 2.3).

Hint
Think back to tolerance limits and how they affect survival of organisms.

TABLE 2.3 Factors affecting population growth and carrying capacity

Density-dependent factors (effect intensifies as density increases)	Density-independent factors (effect occurs regardless of population size)
• Competition: food/nutrient availability, space, shelter, reproductive mate • Predation: proximity to predator/prey • Disease: transmission more effective when there is more frequent close contact	• Climate: temperature, humidity, atmospheric pressure, wind, rainfall • Pollution: amount and type, e.g. toxic runoff or gases released into the atmosphere • Natural disasters: floods, cyclones, tornadoes, volcanic eruptions, earthquakes, tsunamis, storms, bushfire • Extreme climatic events: drought, global warming

Hint
Can you link population growth to r-selected or K-selected species? Which is more likely to experience exponential growth?

2.2.2 Carrying capacity

Carrying capacity is the size of population that can be supported indefinitely on the available resources and services of that ecosystem. Often, intraspecific competition leads to the population reaching carrying capacity. Carrying capacity is a characteristic of logistic growth and occurs when (birth rate + immigration rate) = (death rate + emigration rate) and indicates a stable population, although fluctuations may occur. If the birth rate is significantly more than the death rate, then the growth is still exponential. If the death rate is significantly more than the birth rate, then the population is collapsing. The factors listed in Table 2.3 affect the survival and reproductive success of individuals in a population; hence, they affect the carrying capacity of any environment.

2.3 Changing ecosystems

2.3.1 Succession

The abiotic and biotic components of every ecosystem are constantly fluctuating. Changes can be large or small.

An **ecological succession** is a landscape's response to a disturbance that results in predictable and orderly changes to the structure and composition of the ecosystem. A primary succession is the growth and development of plants on a previously bare or soil-free substrate. A secondary succession occurs where there has been previous growth and soil still exists. Major disturbances include volcanic eruption, dune formation and glacial retreat. Disturbances that can change the availability of resources, and lead to secondary succession, include storms, bushfires, floods, drought, landslides and tree falls.

The succession begins with a **pioneer** community, then transitions through successional stages called **seres**, forming seral communities. The final stable, mature and more complex stage is called the **climax** community. Productivity, diversity and food web complexity all increase as the succession progresses. Primary and secondary successions are summarised in Figures 2.16–2.18 and Table 2.4.

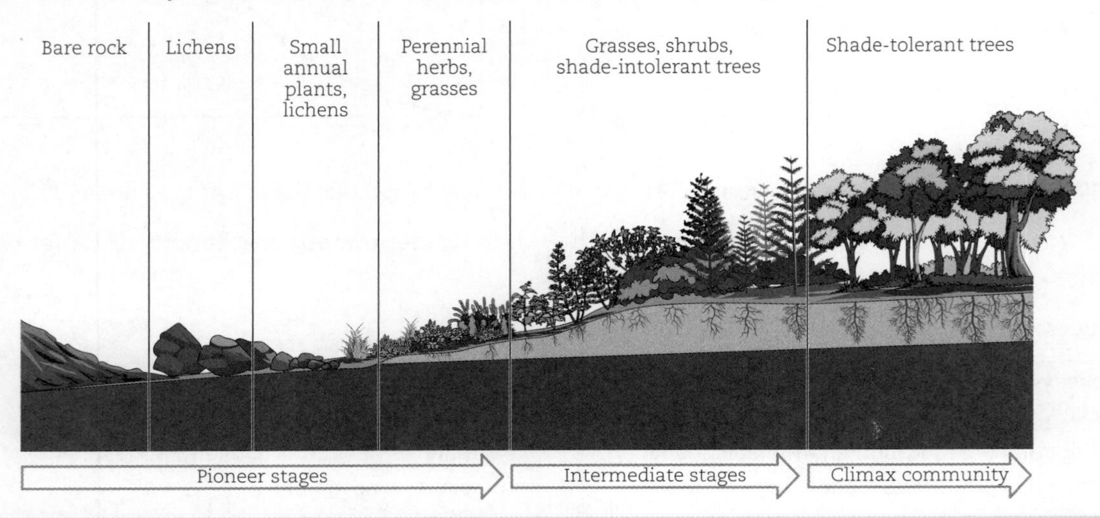

FIGURE 2.16 Stages of a primary succession: from the pioneer lichen species to the seral stage that includes annual and perennial plants with roots, to the climax community of trees

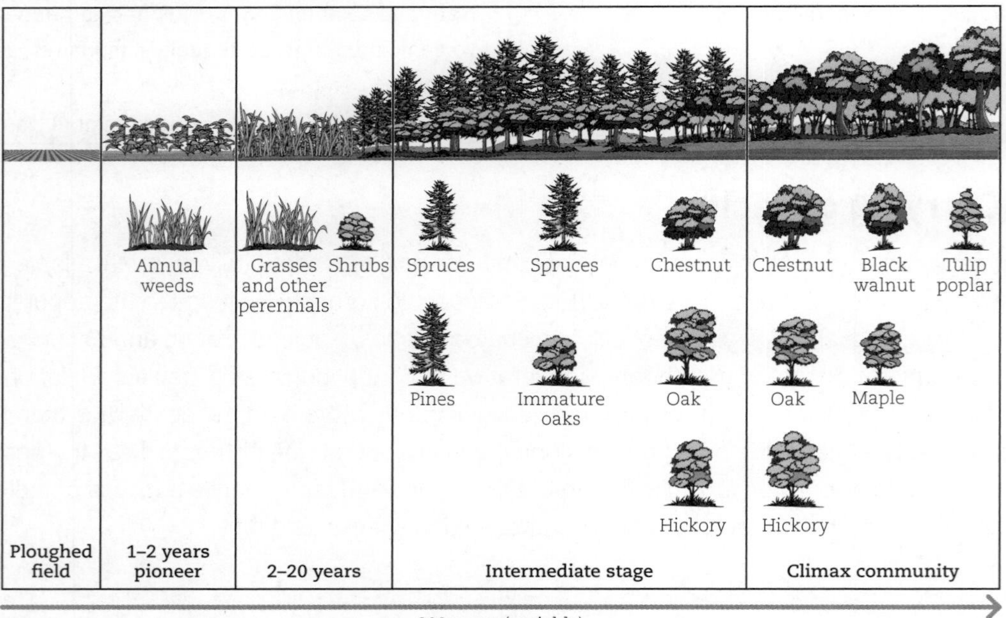

FIGURE 2.17 Stages of a secondary succession: from the pioneer plant species to the seral stage that includes annual and perennial plants with roots, to the climax community of trees

9780170459136

Primary
- Sand dune, glacial retreat, volcano
- No previous growth
- Bare rock
- Starts with lichen
- Biomass low
- Relatively slow

Primary and secondary
- Pioneer species
- Gradual
- Climax community
- Seres

Secondary
- Soil exists
- Previously existing life
- After a natural disaster, e.g. flood, fire, cyclone
- Biomass high
- Relatively fast

FIGURE 2.18 Differentiating between primary and secondary succession

TABLE 2.4 Summary of attributes of species at different stages of a succession

Attributes of pioneer species	Attributes of climax community species
Generally r-selected (rapidly reach reproductive maturity, high number of offspring)	Generally K-selected
Wide tolerance limits	Shade tolerant
High seed dispersal rates	Able to survive in a resource-scarce environment
Dormancy	
Require higher light intensity	
Nitrogen fixers	

2.3.2 Analysis of data

Temporal successional change

In 1980, Mount St Helens erupted on the west coast of the United States, flattening 518 km^2 of land around it. The sequence of photos in Figure 2.19 shows the changes over time (**temporal** changes) since the 1980 eruption. By looking at the number and types of species present, it is possible to predict which stage of a succession the community is.

a

iStock.com/Jeff Goulden

b

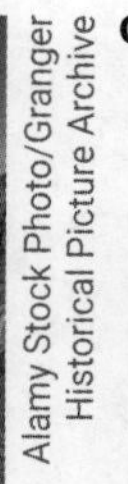

Alamy Stock Photo/Granger Historical Picture Archive

c

Alamy Stock Photo/Terry Donnelly

FIGURE 2.19 Mt St Helens in: a 1980, before eruption; b 1980, after eruption; and c 2009

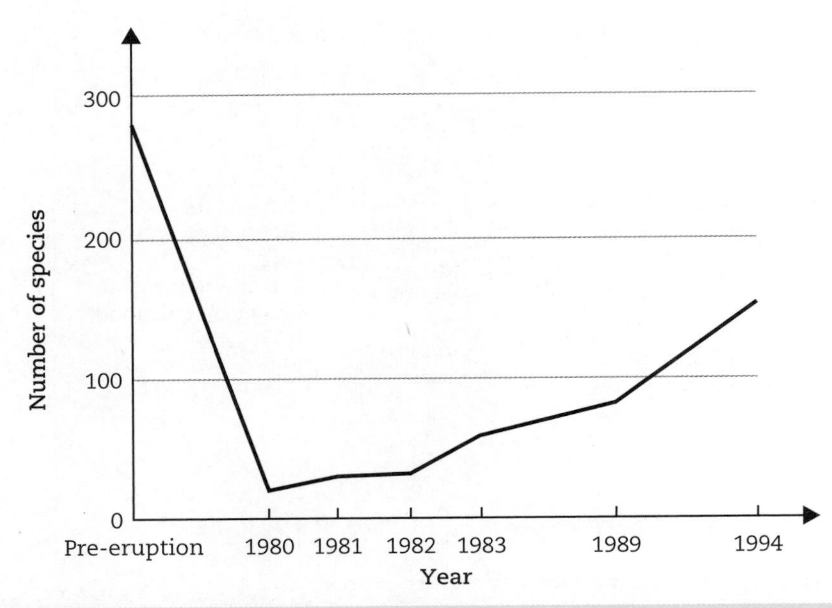

FIGURE 2.20 Temporal change in species number over time at the site of the Mt St Helens volcanic eruption. The data suggests that the area has not yet returned to the climax community before the eruption in 1980. (*x*-axis not to scale.)

Spatial successional change

Investigating the **spatial** distribution of organisms can also provide information about a succession.

Shutterstock.com/nito

FIGURE 2.21 Different parts of a sand dune system represent different stages of a succession. If the dunes pictured are moving towards the sea at 1 metre per year, 50-year-old dunes will be found 50 metres inland.

Fossil data

Our understanding of current ecosystems can be applied to evidence in the fossil record. The fossil record includes any remains or traces of once-living organisms (e.g. grasses and herbivores). Combining this fossil information with data on physical and/or chemical evidence from the same environment as the fossils allows us to draw reasonable conclusions about historical changes ecosystems.

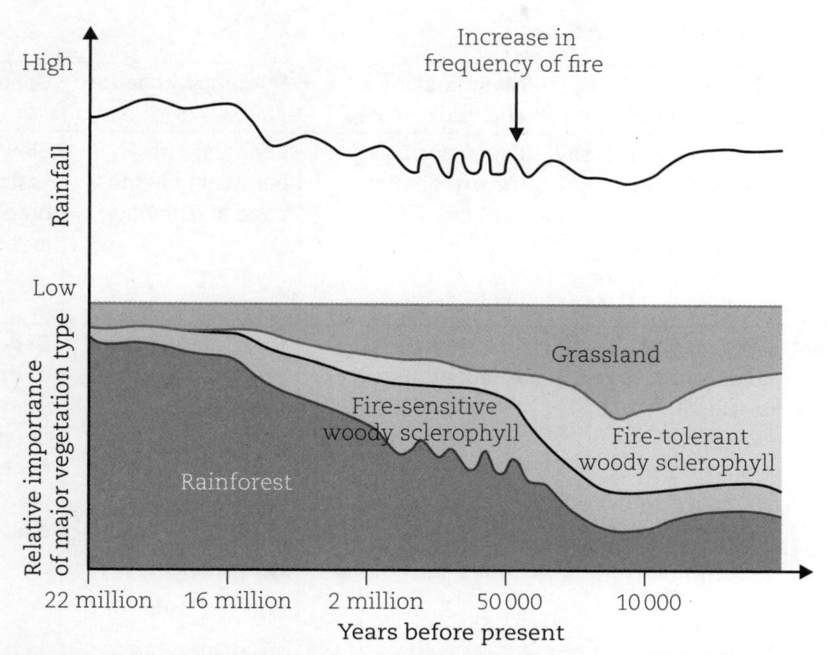

FIGURE 2.22 Changes in Australian vegetation over time. With a decrease in rainfall and a cycle of recent ice ages, the vegetation of Australia has changed dramatically over the last 22 million years.

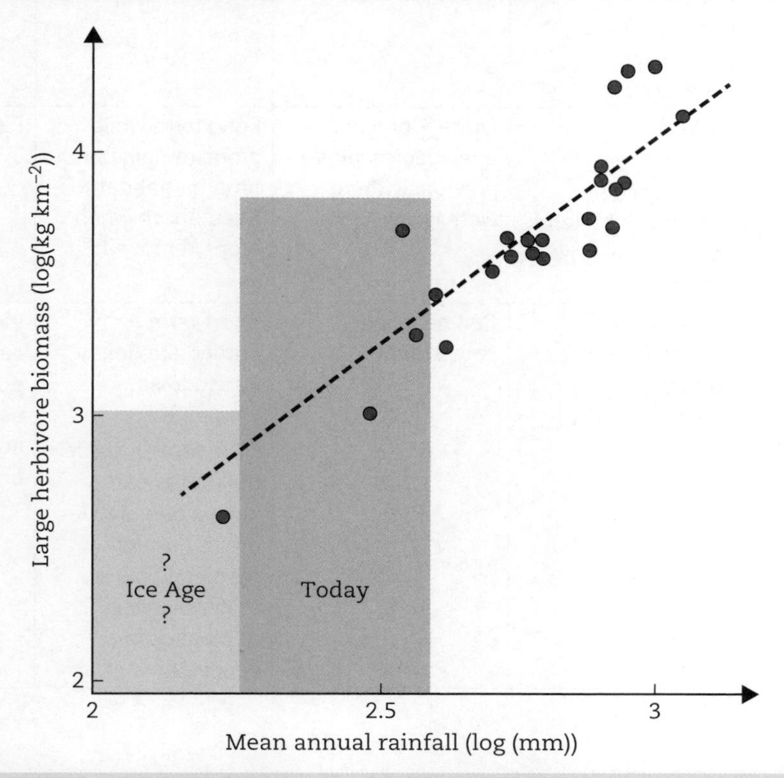

FIGURE 2.23 The total biomass of herbivores in Australia today compared to the biomass of the megafauna of the last ice age. Mean annual rainfall in interior regions of Australia today is shown in the shaded region on the right. Much of the Australian continent became hyper-arid during the coldest times of ice ages (on the left). The total biomass of herbivores was less, probably because of the lower rainfall. Megafaunal herbivores need large amounts of food, and producers were limited by lower rainfall.

2.3.3 Human impact

Although ecosystems are constantly changing, human activities have had a significant effect on the size, duration and speed of ecosystem changes, as well as altering biodiversity.

Table 2.5 and Figure 2.24 summarise the effects of human activity on ecosystems, including species extinction.

TABLE 2.5 Summary of impact of human activities

Activity	Effect on biodiversity	Magnitude of change	Duration of change	Speed of change
Overexploitation: harvesting from a population faster than the natural population can recover	Reduce biodiversity and can lead to extinction of species, e.g. passenger pigeon, dodo	Increasing as the exploitation continues	Long term or permanent in the case of extinction	Slow at first, then faster as the breeding population is reduced
Habitat destruction: a natural environment can no longer support the species present	Reduces biodiversity as areas of natural vegetation are removed and replaced with crops (see monocultures), mines or urban environments Also leads to habitat fragmentation The presence of dams alters freshwater ecosystems	Large – loss of a habitat means the loss of all the species in that area	Short or long term Can lead to succession Changes nutrient cycling, water cycling as well as the number and type of species	Depending on scale of destruction – small-scale destruction generally has a slower overall change
Monocultures: cultivation of a single crop, variety or breed at one time	Reduces biodiversity Approximately 24% of Earth's land surfaces are now cropland	Large – growing one species where previously there were many	Long term while crops remain plus time for habitat to re-establish when crops removed	Fast
Pollution: introduction of substances (e.g. chemicals, plastics) or noise, light or heat, to the environment	Can lead to a change in species composition and loss of diversity	Can be locally catastrophic	Short term – immediate death due to toxin presence Long term – some pollutants can have a cumulative effect through biomagnification or persistence of plastics in an ecosystem	Increasing, especially with pollutants that accumulate either in the environment or in organisms.

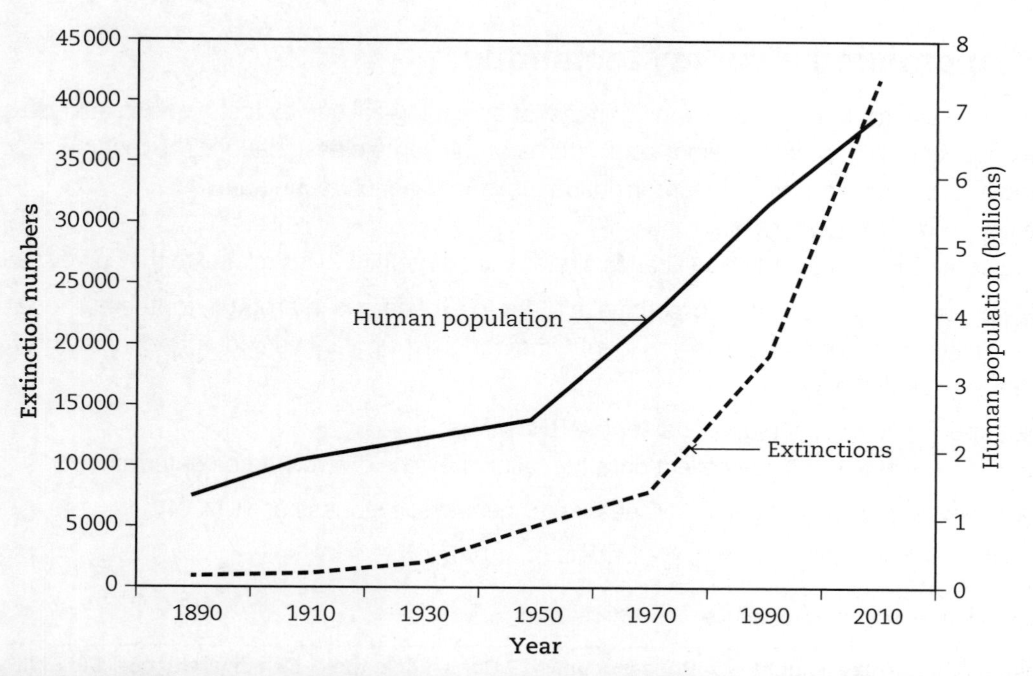

FIGURE 2.24 The growth of the human population compared with extinction numbers for other species

In Australia, pressures affecting biodiversity include:

- extreme weather events
- clearing and fragmentation of habitats
- invasive plants and animals, marine debris and micropollutants
- broad-scale diffuse pollution
- grazing
- fire regimes
- river regulation.

2.3.4 Appraising a survey technique

The syllabus asks for the selection and appraisal of an ecological survey technique to analyse species diversity between two spatially variant ecosystems of the same classification. The techniques covered are listed in Table 2.6, as well as the information provided by each technique.

The key verbs are 'appraise' and 'analyse'.

- Appraise: evaluate the worth, significance or status of something. This relates to the survey technique.
- Analyse: examine or consider something in order to explain and interpret it, for the purpose of finding meaning or relationships and identifying patterns, similarities and differences. This relates to the species diversity.

When appraising a technique, consider the following:

- Does the selected technique collect data that allows for the diversity to be determined?
- How is diversity being defined and measured – richness, evenness or SDI?
- Should the sampling be random, systematic or stratified?

TABLE 2.6 Summarising ways to appraise survey techniques

Technique	Does it count the type of species?	Does it count the number of individuals of each species?	Does it give the distribution of species?	Can diversity be determined?	Suitability for habitat/species
Quadrat	Yes	Yes	No	SDI Evenness Richness	Decide on the basis of what is being sampled, e.g. trees or grass or shrubs.
Line transect	Yes	No	Yes	Richness	
Belt transect	Yes	Yes	Yes	SDI Evenness Richness	
Traps/nets	Yes	Yes	No	SDI Evenness Richness	
Capture–mark–recapture	Only one species chosen	Yes	No	Population estimate for one species only	

Once the data is collected and analysed, the technique used is evaluated to decide whether it was a suitable choice.

Glossary

A+ DIGITAL FLASHCARDS

Revise this topic's key terms and concepts by scanning the QR code or typing the URL into your browser.

https://get.ga/aplus-qce-bio-u34

Autotroph
An organism that can produce its own food from inorganic chemicals (e.g. carbon dioxide, water) and an energy source such as light.

Carrying capacity
The size of the population that can be supported indefinitely on the available resources and services of that ecosystem.

Climax
The final stage of an ecological succession as determined by the available resources and services of that ecosystem.

Combustion
The reaction of a fuel with oxygen to produce energy.

Condensation
The process by which water vapour changes to liquid, releasing energy.

Consumer
An organism that is a heterotroph.

Competitive exclusion principle
Two species cannot occupy the same niche indefinitely.

Decomposer
A heterotroph that feeds on dead or decaying matter.

Denitrifying
The process in which micro-organisms convert nitrogen compounds in the soil into gaseous nitrogen compounds, such as nitrous oxide and nitrogen.

Density dependent
A process in which population growth rates are regulated by past and present population size.

Density independent
A process that affects growth rate regardless of population size or density, e.g. natural disasters.

Detritovore
A heterotroph that feeds on detritus, plant and animal parts.

Ecological pyramid
A graphical representation of the trophic levels in a food chain, showing the number, biomass or energy stored at each level.

Ecological succession
A process of change in the species structure of an ecosystem over time.

Energy efficiency
The proportion of energy that is passed from one trophic level to the next compared to the total energy available.

Evaporation
The process by which water changes from liquid to water vapour, using energy.

Exponential
Growth in which the number of individuals in a population increases at a faster and faster rate.

Food chain
One path that energy and material can flow through in a habitat, beginning with producers.

Food web
The sum of all food chains in an ecosystem.

Gross primary productivity
The total carbon fixed by photosynthesis in all producers (plants) in an ecosystem.

Heterotroph
An organism that gains energy by eating other plants and animals.

Keystone species
A plant or an animal that has a unique and crucial role in the way an ecosystem functions.

Logistic
Growth in which the rate of increase slows as the population approaches the carrying capacity of its environment.

Net primary productivity
The GPP minus energy lost to metabolism and maintenance, i.e. the energy stored as biomass in all primary producers in an ecosystem.

Niche
The role and space that an organism occupies in an ecosystem, including all its interactions with the biotic and abiotic factors of its environment.

Nitrifying
The process in which microorganisms convert gaseous ammonia into nitrites and nitrates.

9780170459136

Nitrogen fixation
The process in which micro-organisms remove nitrogen from the atmosphere and convert it to biologically useful nitrogen compounds.

Photosynthesis
The process in which plants convert carbon dioxide and water into carbohydrates and oxygen in the presence of light energy and chlorophyll.

Pioneer
A species that is first to colonise a barren or disrupted environment.

Population
A group of organisms of the same species that live and breed in a given area.

Precipitation
Liquid or solid water (e.g. rain, hail), formed in the atmosphere and returning to Earth.

Producer
An organism that is an autotroph.

Respiration
The process in cells in which oxygen reacts with glucose to produce energy stored as ATP.

Seres
An intermediate stage in an ecological succession.

Sink (carbon)
An accumulation of carbon-containing material that removes carbon dioxide from the atmosphere for an indefinite period.

Source (carbon)
Any organic or inorganic compound containing carbon that an organism can use to synthesise the organic compounds it needs to live.

Spatial
The distribution of a species in space within an ecosystem.

Temporal
The distribution of species in an ecosystem through time.

Transfer (energy)
The passage of energy through a food chain.

Transform (energy)
The changes in energy as it passes through an ecosystem from light energy (from the Sun) to chemical energy in plants, herbivores, and carnivores, to mechanical energy of movement.

Transpiration
The loss of water from a plant by evaporation at the leaf surface.

Trophic level
A group of organisms occupying the same level in a food chain, e.g. herbivores.

9780170459136

Revision summary

Use the following summary of syllabus dot points and key knowledge within Unit 3 Topic 2 to ensure that you have thoroughly reviewed the content. Provide a brief definition or comment for each item to demonstrate your understanding or code them using the traffic light system – green (all good), amber (needs some review), red (priority area to review).

Functioning ecosystems	
• sequence and explain the transfer and transformation of solar energy into biomass as it flows through biotic components of an ecosystem, including – converting light to chemical energy – producing biomass and interacting with components of the carbon cycle	
• analyse and calculate energy transfer (food chains, webs and pyramids) and transformations within ecosystems, including – loss of energy through radiation, reflection and absorption – efficiencies of energy transfer from one trophic level to another – biomass	
• construct and analyse simple energy-flow diagrams illustrating the movement of energy through ecosystems, including the productivity (gross and net) of the various trophic levels	
• describe the transfer and transformation of matter as it cycles through ecosystems (water, carbon and nitrogen)	
• define *ecological niche* in terms of habitat, feeding relationships and interactions with other species	
• understand the competitive exclusion principle	

››

››	
• analyse data to identify species (including microorganisms) or populations occupying an ecological niche	
• define *keystone species* and understand the critical role they play in maintaining the structure of a community	
• analyse data (from an Australian ecosystem) to identify a keystone species and predict the outcomes of removing the species from an ecosystem.	
Population ecology	
• define the term *carrying capacity*	
• explain why the carrying capacity of a population is determined by limiting factors (biotic and abiotic)	
• calculate population growth rate and change (using birth, death, immigration and emigration data)	
• use the Lincoln Index to estimate population size from secondary or primary data	
• analyse population growth data to determine the mode (exponential growth J-curve, logistic growth S-curve) of population growth	

››

›› • discuss the effect of changes within population-limiting factors on the carrying capacity of the ecosystem.	
Changing ecosystems	
• explain the concept of ecological succession (refer to pioneer and climax communities and seres)	
• differentiate between the two main modes of succession: primary and secondary	
• identify the features of pioneer species (ability to fixate nitrogen, tolerance to extreme conditions, rapid germination of seeds, ability to photosynthesise) that make them effective colonisers	
• analyse data from the fossil record to observe past ecosystems and changes in biotic and abiotic components	
• analyse ecological data to predict temporal and spatial successional changes	
• predict the impact of human activity on the reduction of biodiversity and on the magnitude, duration and speed of ecosystem change.	
• Mandatory practical: Select and appraise an ecological surveying technique to analyse species diversity between two spatially variant ecosystems of the same classification (e.g. a disturbed and undisturbed dry sclerophyll forest).	

Exam practice

Multiple-choice questions

Each multiple-choice question is worth 1 mark.

Solutions start on page 155.

Question 1

Which image best represents competitive exclusion of two protozoan species?

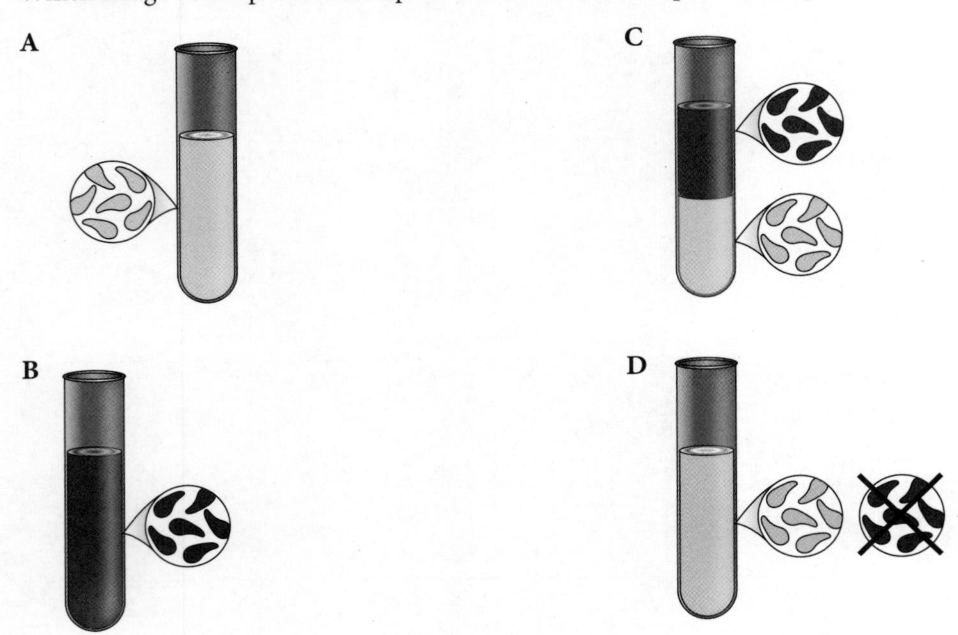

The following diagram relates to Questions 2–5.

The diagram shows a food web that includes the northern quoll in Kakadu National Park.

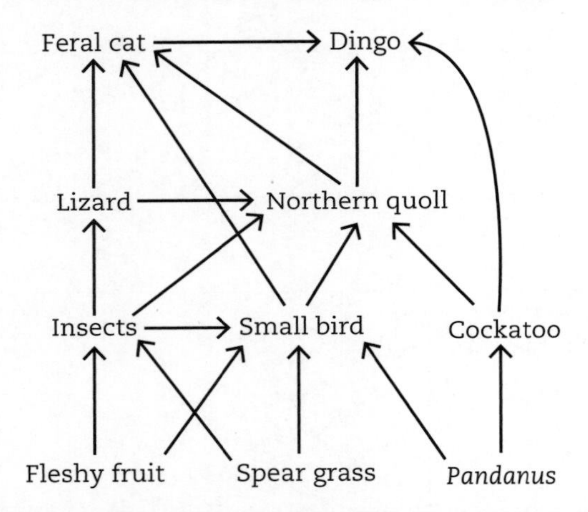

Kakadu National Park food web

Question 2

A food chain that finishes with a secondary consumer is

A *Pandanus* → cockatoo → dingo.

B Spear grass → insects → lizards → feral cat.

C Fleshy fruit → small bird → northern quoll → dingo.

D Spear grass → insects → small bird → northern quoll → dingo.

Question 3

The northern quoll occupies

A trophic level 3 only.

B trophic levels 2 and 3.

C trophic levels 3 and 4.

D trophic levels 2, 3 and 4.

Question 4

Pandanus is considered

A an autotroph.

B a decomposer.

C a detritovore.

D a heterotroph.

Question 5

The longest food chain in this food web has

A three trophic levels.

B four trophic levels.

C five trophic levels.

D six trophic levels.

Question 6 ©QCAA QCAA 2020 P1 MC Q7

When predicting successional change, which of the following would typically indicate that an ecosystem is progressing towards its climax community?

	Abundance of K-selected species	Biomass
A	Increasing	Decreasing
B	Decreasing	Increasing
C	Increasing	Increasing
D	Decreasing	Decreasing

Short-response questions

Solutions start on page 156.

Question 7 (5 marks)

Sequence and explain what happens to light energy after it is absorbed by leaves of the *Pandanus* in the Kakadu food web.

Question 8 (3 marks)

In an aquatic ecosystem in Silver Springs, Florida, USA, the net productivities – rates of energy storage as biomass (in kJ m^{-2} $year^{-1}$) for trophic levels were as follows:

- Primary producers, such as plants and algae: 31 996
- Primary consumers, such as snails and insect larvae: 4633
- Secondary consumers, such as fish and large insects: 446
- Tertiary consumers, such as large fish and snakes: 21

a Calculate the energy efficiency for the transfer of energy from trophic level 2 to trophic level 3 to one decimal place.

$$\text{efficiency} = \frac{\text{energy transferred to next level}}{\text{total energy in}} \times 100$$

2 marks

b Explain why the energy efficiency is less than 100%. 1 mark

Question 9 (1 mark)

A kangaroo has eaten 10 kg of biomass in the form of grass and lost a total of 6.3 kg in the form of faeces (waste), urine and gas (excretion). The increase in mass of its body tissues is 0.4 kg. Determine how much biomass has been used up in respiration.

Question 10 (2 marks)

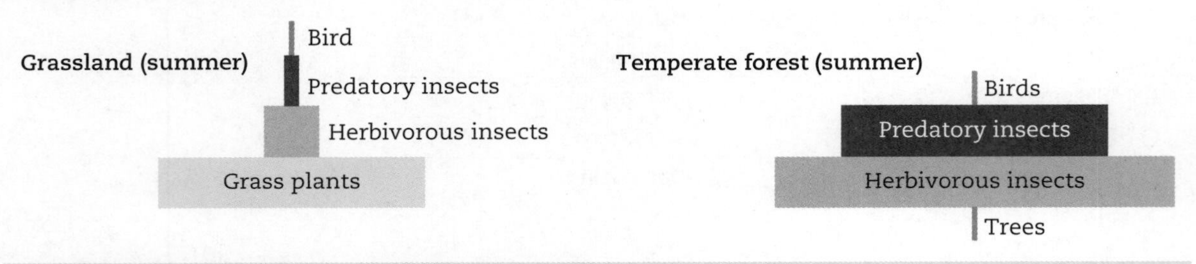

Pyramids of numbers for a grassland and a forest

Explain why the pyramids of energy for grassland and forest would look the same when the pyramids of number are different.

Question 11 (10 marks)

a In what type of succession would you expect to see lichen on a rock? Provide a reason for your answer. 2 marks

b The photo shows an abandoned dirt road in south-east Queensland. Predict how you would expect this road to change over time. 3 marks

Attribute of succession	Now	In 50 years
r-selected or K-selected species?		
Biodiversity low or high?		
Biomass low or high?		

c Provide an explanation for your predictions in part **b**. 2 marks

d List three attributes you would expect pioneer species to have to successfully colonise this abandoned road. 3 marks

Question 12 (5 marks)

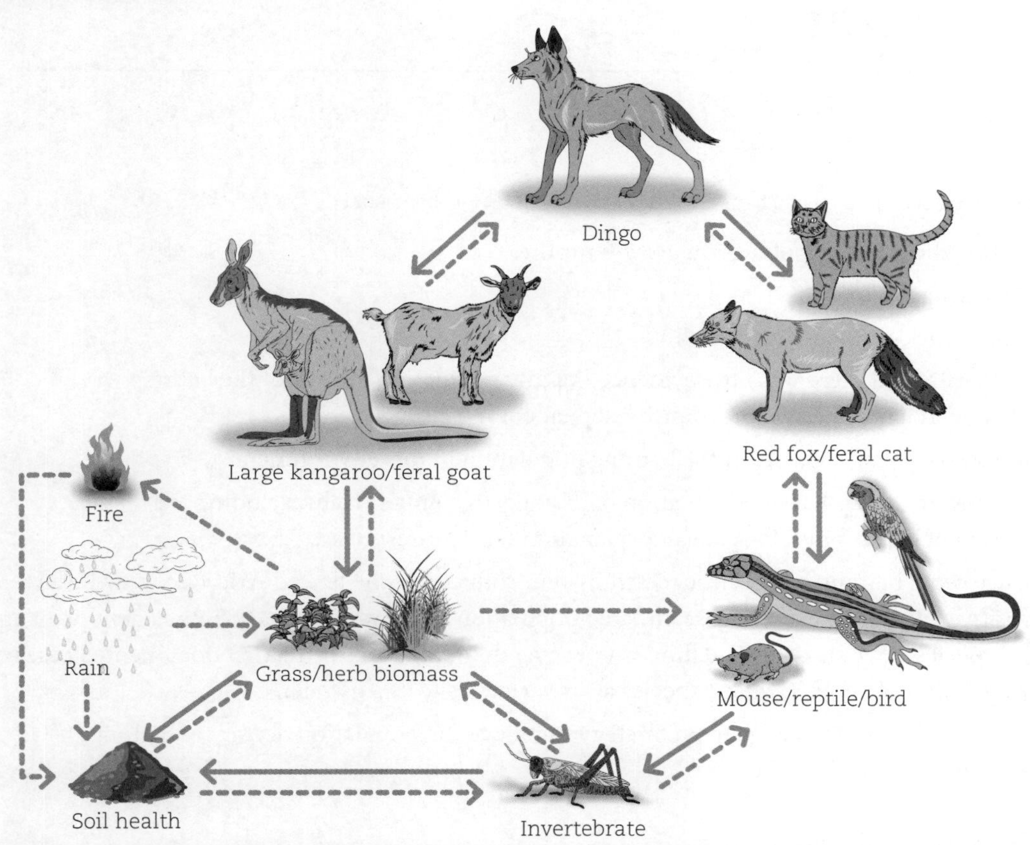

Potential effects of re-introducing dingoes to Australian ecosystems. A dingo-induced trophic cascade (solid arrows), where the presence of dingoes has both direct and indirect effects. Researchers suggest that reintroducing dingoes can have a positive effect. Dashed arrows represent the food web.

a Define 'keystone species'. 1 mark

b Explain how the introduction of dingoes could affect the biomass of grasses and herbs. 2 marks

c Use evidence from the diagram to show how this food web is linked to the carbon cycle and the water cycle. 2 marks

Question 13 (7 marks)

The burrowing bettong (*Bettongia lesueur*) is a threatened marsupial reintroduced to a fenced reserve in arid Australia.

Activities of another two reintroduced species, the greater stick-nest rat (*Leporillus conditor*) and the bandicoot, were also recorded. The bettong and the rat are both herbivores. The bandicoot feeds on insects and grubs.

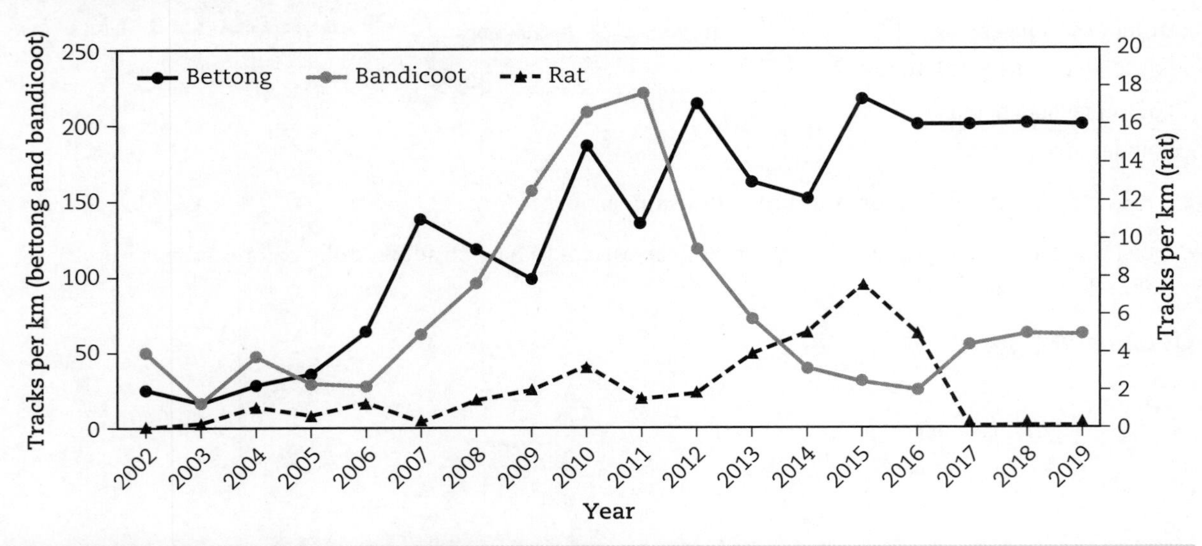

Changing populations of burrowing bettong, greater stick-nest rat and bandicoot

a Determine the mode of population growth for the: 2 marks

i burrowing bettong

ii greater stick-nest rat.

b Researchers found there was little evidence that the population growth of the bettong was density dependent. Explain what density dependent means. 1 mark

c Suggest the carrying capacity of the bettong population in this environment. 1 mark

d Is there any evidence that the population of bettongs is competitively excluding the population of bandicoots? Provide an explanation for your response. 2 marks

e The burrowing bettong has been successfully reintroduced to the fenced Arid Recovery Reserve, but the positive average rate of increase, inflated population density and impacts on resident plant and animal species suggests the population is now overabundant. This is the first documented case of overpopulation of a reintroduced species at a restricted site in Australia.

Suggest an appropriate management strategy to reduce the population density of bettongs in the Arid Recovery Reserve. 1 mark

Question 14 (6 marks)

Humans introduced rabbits to Australia in 1859 for hunting. Rabbits were released into the wild in Victoria.

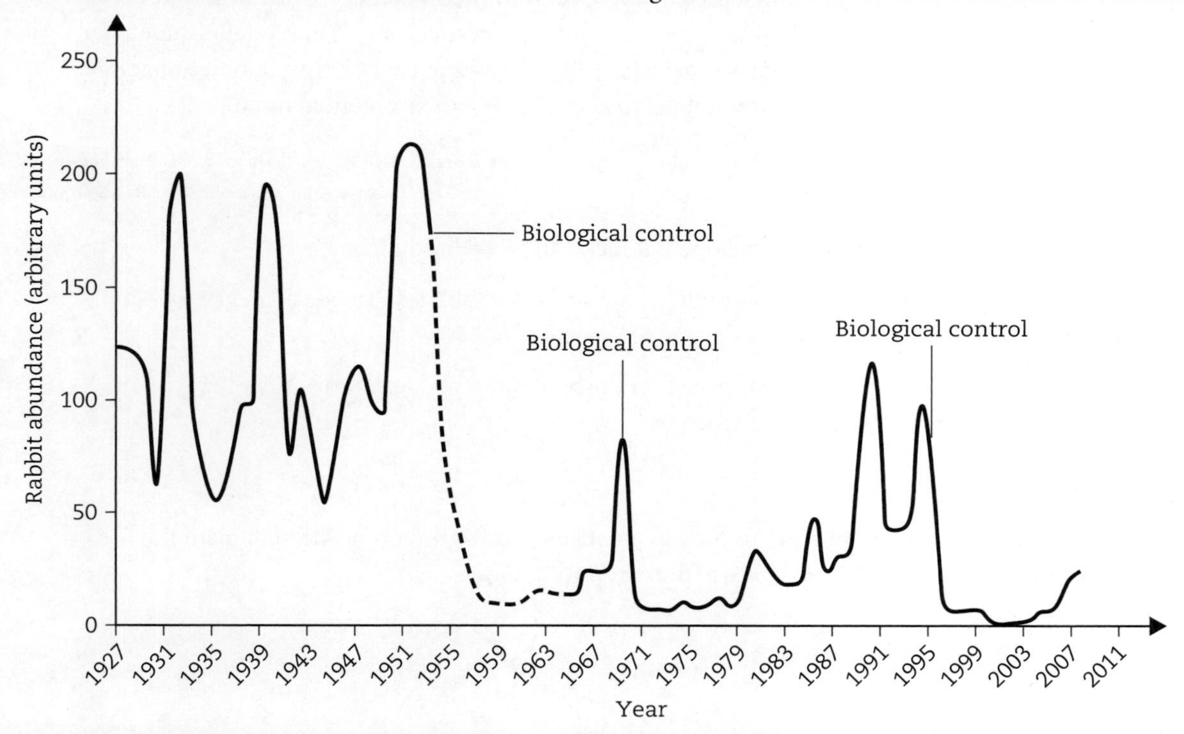

Rabbit abundance in semi-arid Australia in response to biological control actions (the arrows)

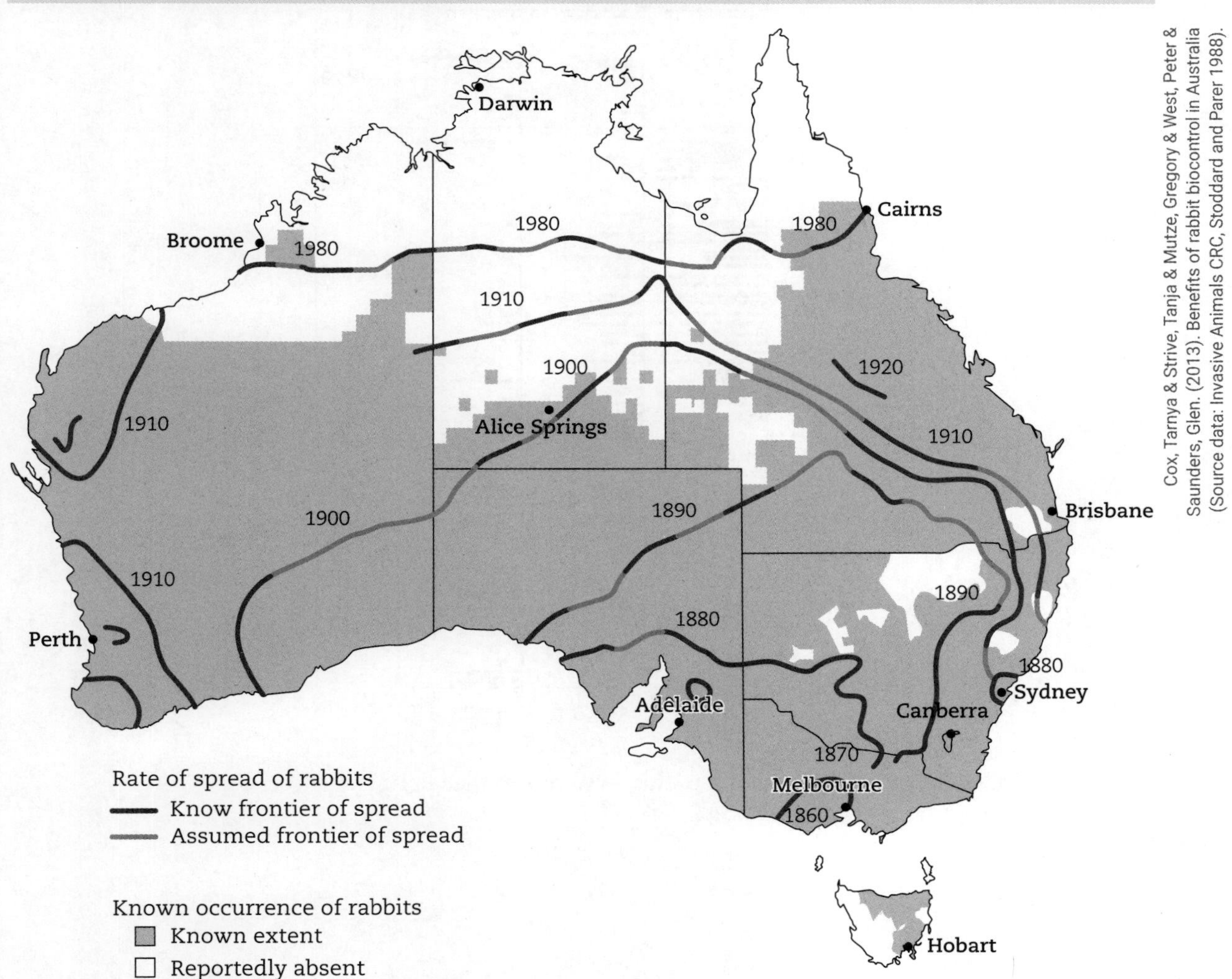

Rate of spread of wild rabbits since their introduction to Geelong, Victoria, in 1859 to their current known distribution

Cox, Tarnya & Strive, Tanja & Mutze, Gregory & West, Peter & Saunders, Glen. (2013). Benefits of rabbit biocontrol in Australia (Source data: Invasive Animals CRC, Stoddard and Parer 1988).

The following extract is from 'Benefits of rabbit biocontrol in Australia'

> The extreme sensitivity of many native plant species to rabbit damage – as few as one rabbit per 10 000 m² can impede natural regeneration – has resulted in 75 nationally threatened plant species and five threatened ecological communities being at risk from rabbit impacts … [Biological control] limits rabbit numbers to about 15% of their potential numbers.

Cox T., Strive T., Mutze G., West P., and Saunders G. (2013). *Benefits of rabbit biocontrol in Australia* (Invasive Animals Cooperative Research Centre)

a Explain the difference between a fundamental niche and a realised niche. 2 marks

b Have the biological control measures reduced the spread of rabbits? Provide evidence from the graph and map. 2 marks

c Explain possible effects of the rabbit spread on the biodiversity of semi-arid Australia had biological control not been implemented. 2 marks

Question 15 (3 marks)

Fossil pollen records of plants at a site in New Zealand show how the composition of plant life has changed from 16 000 calibrated years before present until today.

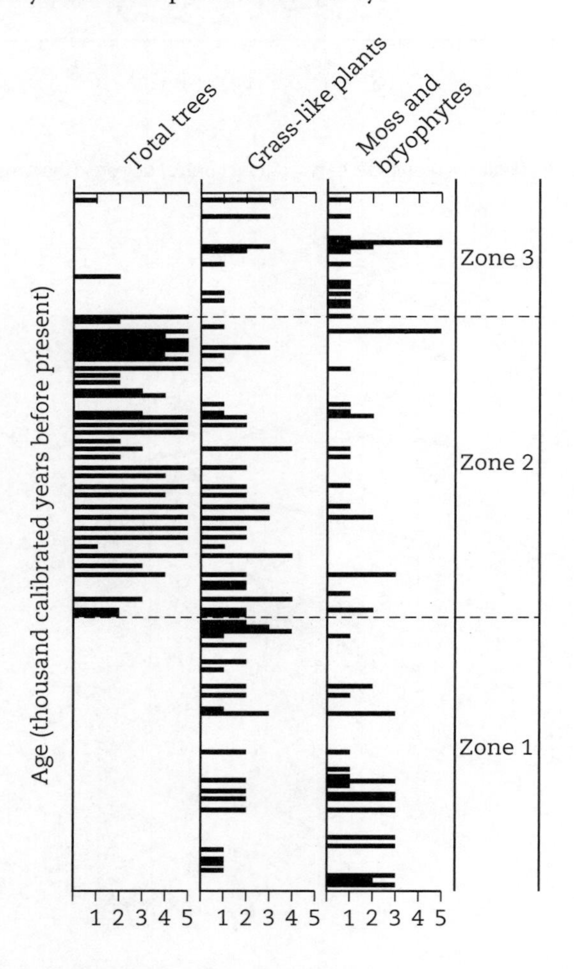

Analyse the data to describe any evidence in this fossil record that a primary succession occurred.

Question 16 (4 marks) ©QCAA QCAA 2020 P1 SR Q28

Identify two differences between primary and secondary ecological succession. Refer to one example of each type of succession in your response.

Question 17 (4 marks) ©QCAA QCAA 2020 P1 SR Q25

The following simplified energy flow diagram provides the gross productivity figures for producers and herbivores in an ecosystem.

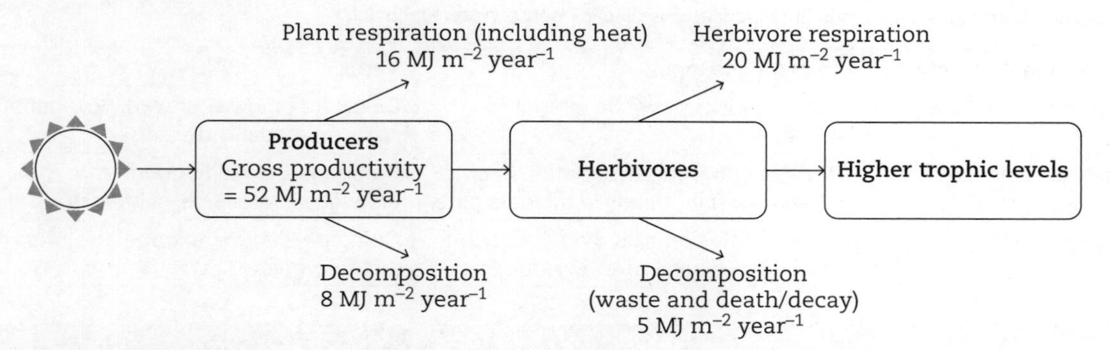

a Determine the net productivity for the producers and herbivores in this ecosystem. 2 marks

Producers:

Net productivity = ______________________ MJ m^{-2} year^{-1}

Herbivores:

Net productivity = ______________________ MJ m^{-2} year^{-1}

b Contrast the outputs of energy for the two trophic levels in the diagram. 2 marks

Question 18 (3 marks) ©QCAA QCAA 2020 P1 SR Q8

In an experiment studying the population dynamics of the house fly, two adult house flies were placed in a fly cage with a limited food supply. Population data was collected, as summarised in the table.

Generation	Number of eggs	Number of larvae	Number of pupae	Number of adults
1	0	0	0	2
2	120	110	95	88
3	250	225	213	210
4	500	475	462	12
5	20	2	0	0

Determine the population growth model exhibited by the house fly. Explain your reasoning.

Chapter 3
Unit 3 Data test

Internal Assessment 1 is a Data test. It addresses Assessment Objectives 2, 3 and 4, requiring you to apply understanding, analyse and interpret evidence. It is completed individually, under supervised conditions with 60 minutes of working time and 10 minutes of perusal time.

TABLE 3.1 Summary of the types of responses possible in Internal Assessment 1

Assessment objectives	The response must:	Verbs
Apply understanding	be an unknown scientific quantity or feature.	Calculate (show your working), identify, recognise, use evidence
Analyse evidence	identify a trend, pattern, relationship, limitation or uncertainty in the data sets.	Categorise, classify, contrast, distinguish, organise, sequence
Interpret evidence	draw a conclusion based on the data sets (not your general knowledge).	Compare, deduce, extrapolate, infer, justify, predict

Hint

Although 'identify' is used in both 'apply' and 'analyse' questions, the expected response is different.

It is important to match the type of response to the verb in the question.

TABLE 3.2 Definitions of cognitive verbs associated with the Data Test Assessment Objectives

Assessment objective	Verb	Definition (QCAA)
Apply	Calculate	Determine or find (e.g. a number, answer) by using mathematical processes; obtain a numerical answer showing the relevant stages of working; ascertain/determine from given facts, figures or information.
	Identify	Distinguish; locate, recognise and name; establish or indicate who or what someone or something is; provide an answer from a number of possibilities; recognise and state a distinguishing fact or figure.
	Recognise	Identify or recall particular features of information from knowledge; identify that an item, characteristic or quality exists; perceive as existing or true; be aware of or acknowledge.
	Use evidence	Operate or put into effect; apply knowledge or rules to put theory into practice.
Analyse	Categorise	Place in or assign to a particular class or group; arrange or order by classes or categories; classify, sort out, sort, separate.
	Classify	Arrange, distribute or order in classes or categories according to shared qualities or characteristics.
	Contrast	Display recognition of differences by deliberate juxtaposition of contrary elements; show how things are different or opposite; give an account of the differences between two or more items or situations, referring to both or all of them throughout.
	Distinguish	Recognise as distinct or different; note points of difference between; discriminate; discern; make clear difference/s between two or more concepts or items.
	Organise	Arrange, order; form as or into a whole consisting of interdependent or coordinated parts, especially for harmonious or united action.
	Sequence	Place in a continuous or connect series; arrange in a particular order.

››

Assessment objective	Verb	Definition (QCAA)
Interpret	Compare	Display recognition of similarities and differences and recognised the significance of these similarities and differences.
	Deduce	Reach a conclusion that is necessarily true, provided a given set of assumptions is true; arrive at, reach or draw a logical conclusion from reasoning and the information given.
	Extrapolate	Infer or estimate by extending or projecting known information; conjecture; infer from what is known; extend the application of something (e.g. a method or conclusion) to an unknown situation by assuming that existing trends will continue or similar methods will be applicable.
	Infer	Derive or conclude something from evidence and reasoning, rather than from explicit statements; listen or read beyond what has been literally expressed; imply or hint at.
	Justify	Give reasons or evidence to support an answer, response or conclusion; show or prove how an argument, statement or conclusion is right or reasonable.
	Predict	Give an expected result of an upcoming action or event; suggest what may happen based on available information.

Data set 1

Solutions start on page 158.

The southern brown bandicoot is a small omnivorous marsupial that lives in scrubby habitats with low ground cover and shelter.

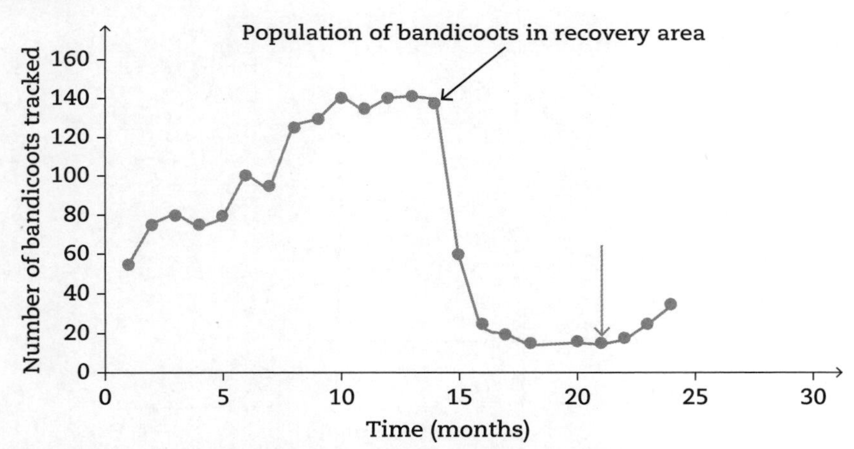

FIGURE 3.1 The southern brown bandicoot population as determined by live trapping

The black arrow pinpoints the entry of a fox to the recovery area. The light grey arrow represents when the fox was found dead in the area.

A capture–mark–recapture sampling process was carried out in month 25 after the death of the fox. The data is summarised in Table 3.3.

TABLE 3.3 Results of the capture–mark–recapture sampling of bandicoots

Original marked sample	Number recaptured	Number recaptured and marked
26	10	4

9780170459136

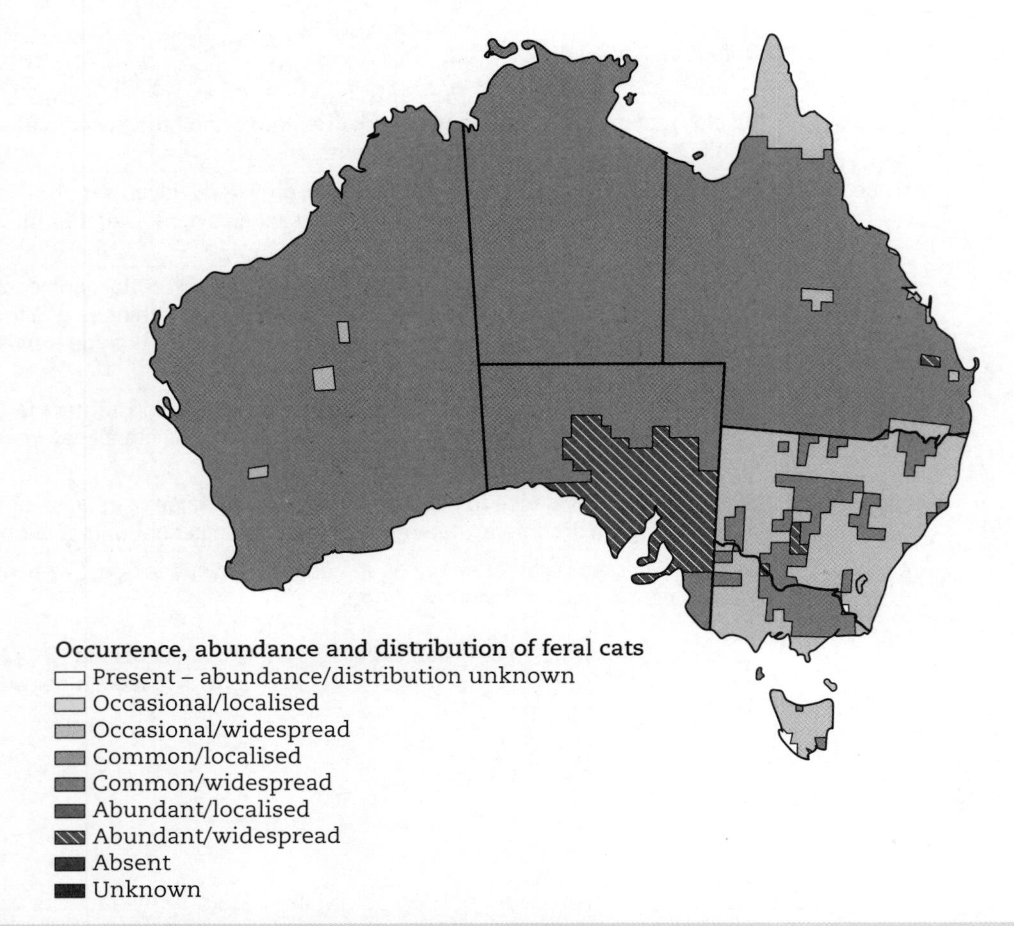

FIGURE 3.2 The abundance and distribution of feral cats in 2017, based on baiting

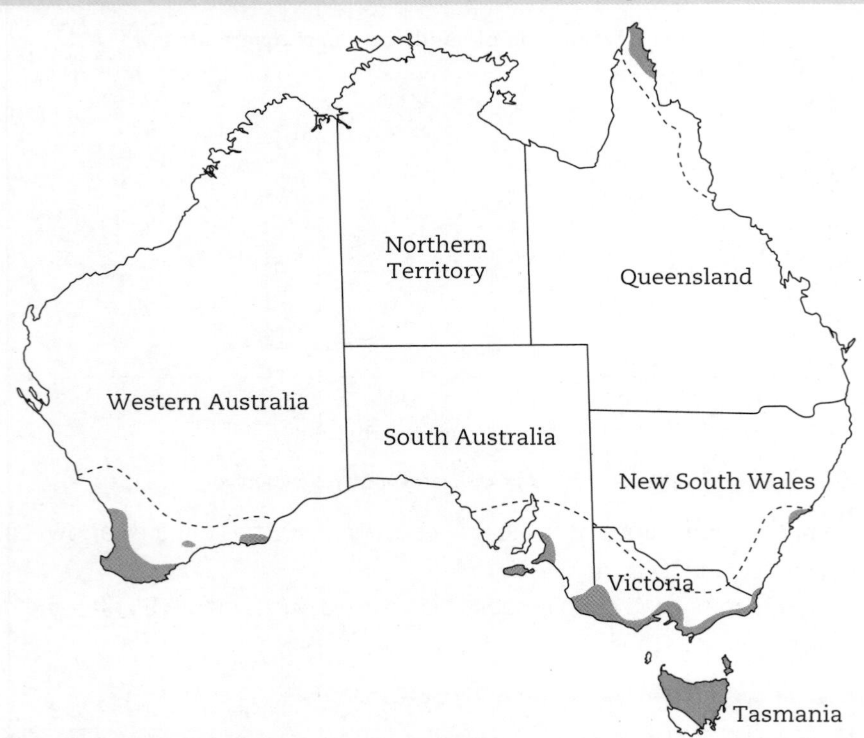

FIGURE 3.3 The current distribution (solid grey) and the historical distribution (dashed line) of the southern brown bandicoot

Lincoln index:

$$N = \frac{M \times n}{m}$$

Question 1: Apply (2 marks)

Calculate the estimated population for month 25, based on the data in Table 3.3 (page 77).

Hint
You will be provided with the rule, but you need to know what the letters represent.

Question 2: Analyse (2 marks)

Classify the population growth of bandicoots before the arrival of the fox and then after the death of the fox.

Question 3: Analyse (1 mark)

Contrast the population distribution of feral cats (Figure 3.2) with the current and historical population distribution of bandicoots (Figure 3.3).

Hint
When figures or diagrams are referred to in the question, you must use information from them in your answer.

Question 4: Interpret (2 marks)

Researchers have concluded that feral cats have an impact on the survival of bandicoots. Justify this conclusion, using information from Figures 3.2 and 3.3.

Question 5: Interpret (2 marks)

Predict what could happen to the bandicoot distribution should the feral cat be removed as a predator, using information from Figures 3.1–3.3.

Data set 2

Solutions start on page 159.

Some researchers decided to investigate plant species found on sand dunes on the Sunshine Coast. Their data is recorded in Table 3.4.

TABLE 3.4 Plants found in five 1 m^2 quadrats placed at site 1 on a sand dune on the Sunshine Coast

Species	Quadrat 1	Quadrat 2	Quadrat 3	Quadrat 4	Quadrat 5	n	Density (m^{-2})
Beach spinifex	2	3	1	2	3	11	2.2
Pigface	3	3	3	1	2	12	2.4
Beach morning glory	1	1	1	1	0	4	0.8
Beach fan flower	1	0	2	1	0	4	0.8

A second site was sampled at a beach 100 metres south of the original site. SDI at this site was 0.47.

TABLE 3.5 Plants found in five quadrats placed at site 2

Species	Quadrat 1	Quadrat 2	Quadrat 3	Quadrat 4	Quadrat 5
Beach spinifex	5	3	4	4	3
Pigface	1	2	1	1	1
Beach morning glory	0	0	0	1	0
Beach fan flower	1	0	0	0	0

Simpson's diversity index:

$$SDI = 1 - \frac{\sum n(n-1)}{N(N-1)}$$

Question 1: Apply (3 marks)

Calculate SDI for site 1.

Hint
You will be provided with the rule, but you need to know what the letters represent.

Question 2: Apply (1 mark)

Identify the most commonly occurring species at site 2.

Question 3: Analyse (1 mark)

Sequence the plants at site 1 from most to least dense.

Question 4: Analyse (1 mark)

Contrast the density of the pigface at both sites.

Hint
It is expected that you will know 'compare' means you need to discuss a similarity, a difference and the significance of these.

Question 5: Interpret (3 marks)

Compare the SDI for sites 1 and 2.

Data set 3

Solutions start on page 160.

Sequencing genomes has proven to be a powerful tool for resolving the relationships of different species across the tree of life.

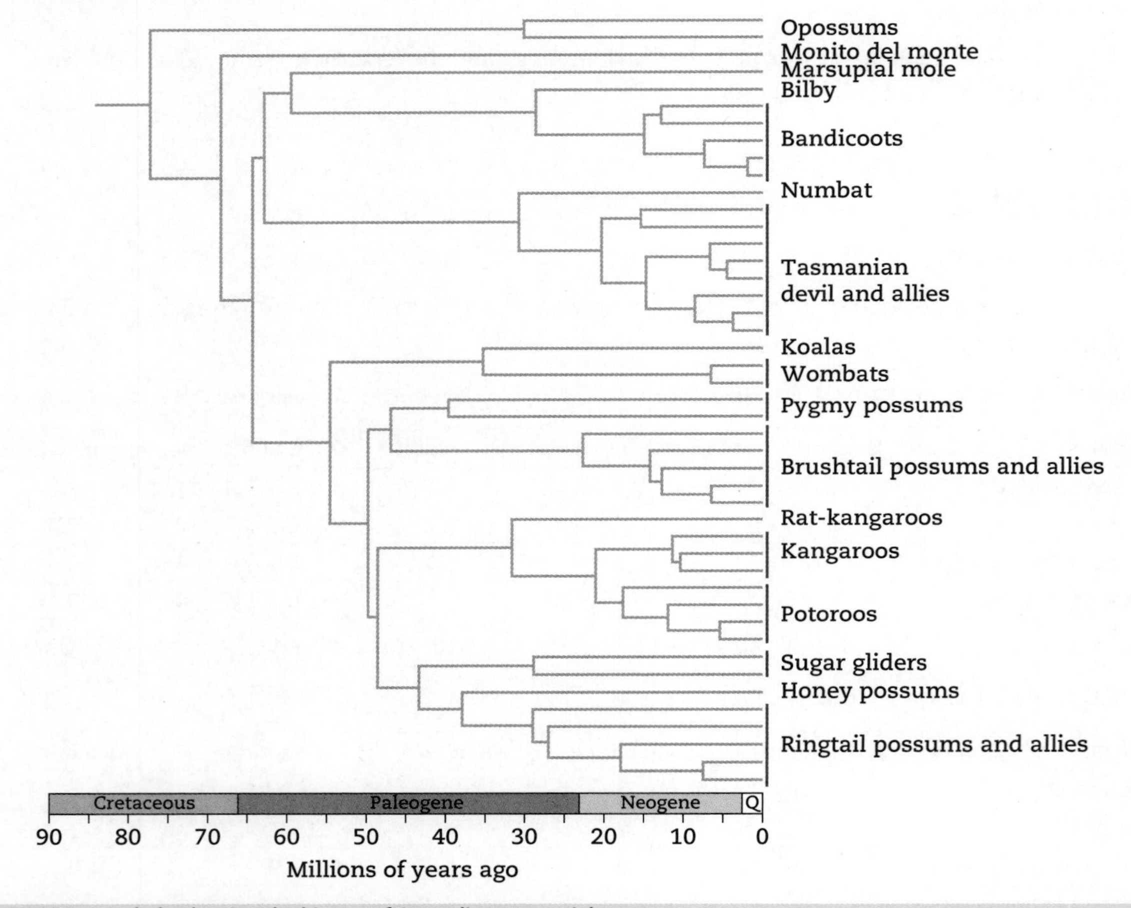

FIGURE 3.4 A clade showing the history of Australian marsupials

Question 1: Apply (1 mark)

Identify the outgroup.

Question 2: Apply (1 mark)

Determine how many million years ago koalas and wombats had a common ancestor.

Question 3: Analyse (1 mark)

Organise these marsupials from most to least related: sugar glider, honey possum, brushtail possum.

Question 4: Interpret (2 marks)

Deduce whether pygmy possums are more closely related to ringtail possums or to Tasmanian devils.

Question 5: Interpret (2 marks)

Predict which animal would be most similar to an 'ancestral marsupial'.

Question 6: Analyse (2 marks)

Contrast the development of ringtail possums with that of pygmy possums.

Question 7: Apply (1 mark)

Identify on the cladogram extract the last common ancestor of honey possums and brushtail possums.

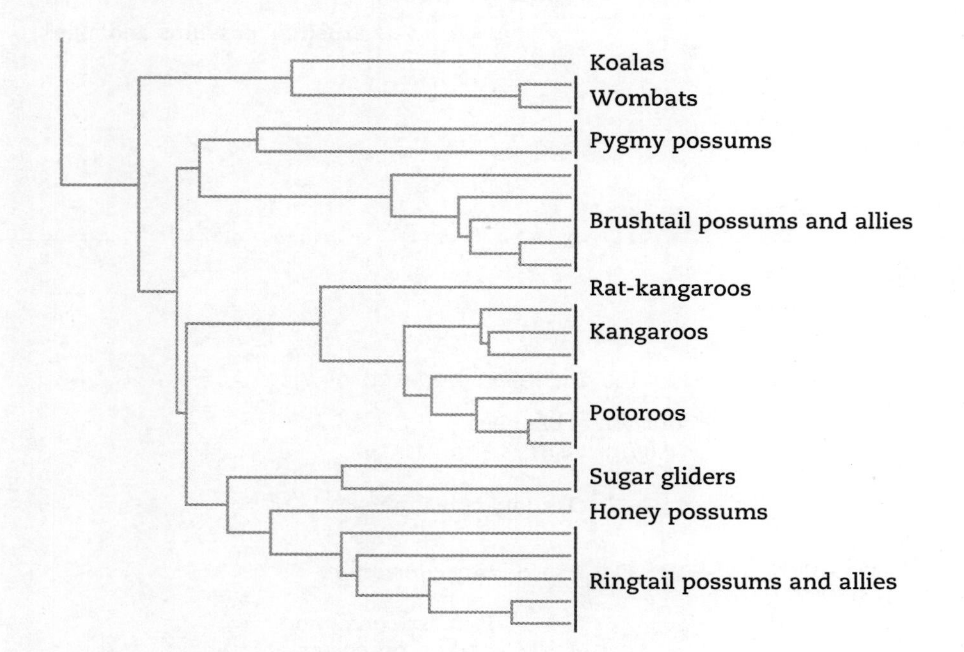

9780170459136

Question 8: Apply (1 mark)

Identify on the cladogram extract the clade that includes both bandicoots and numbats.

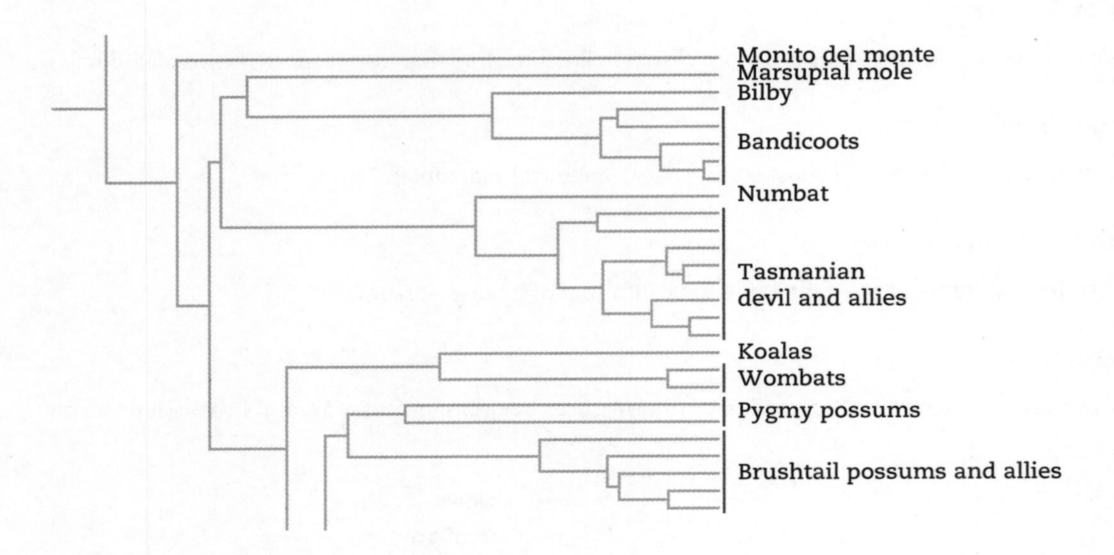

Data set 4

Solutions start on page 161.

Figure 3.5 shows a biomass pyramid for the Australian food web in Figure 3.6.

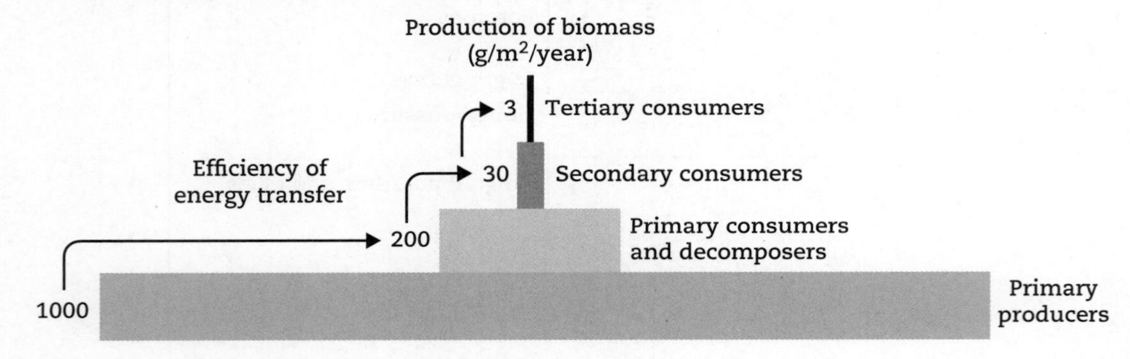

FIGURE 3.5 A biomass pyramid for an Australian food web

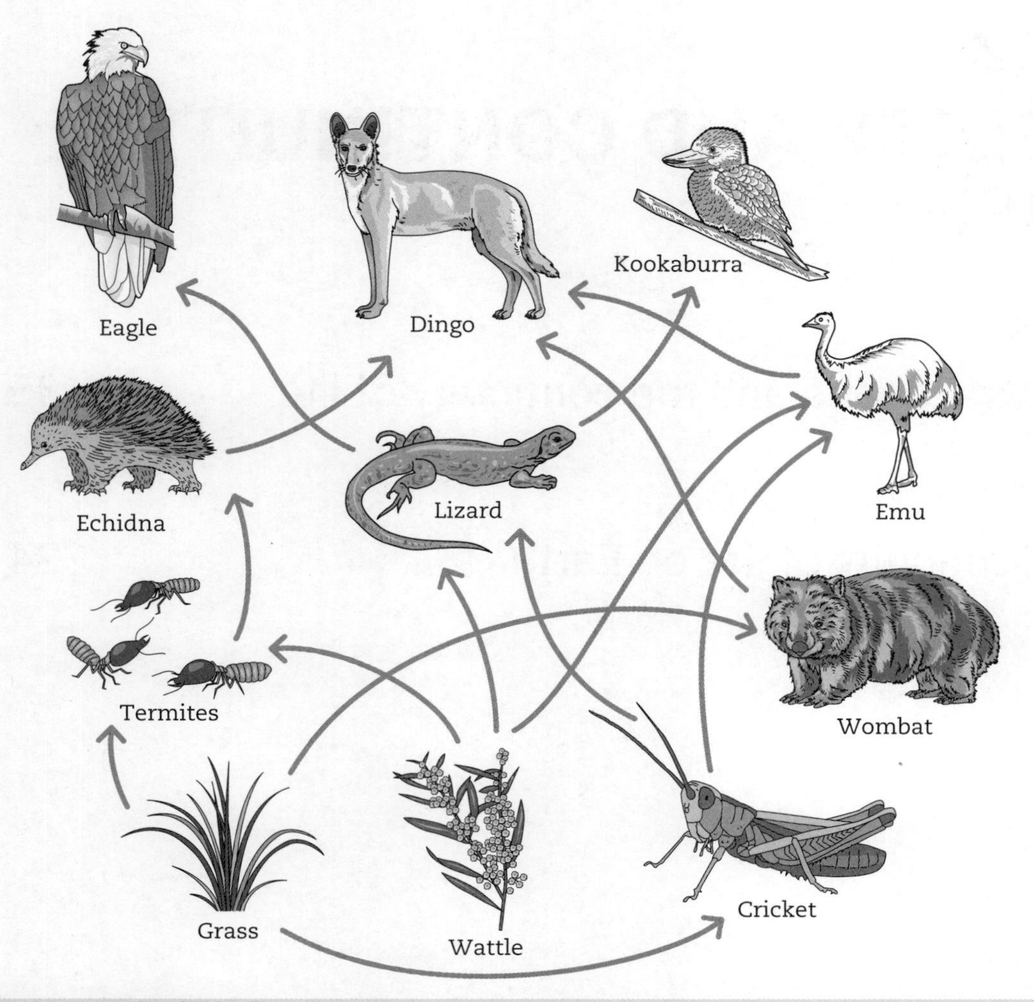

FIGURE 3.6 An Australian food web

Energy efficiency:

$$\text{efficiency} = \frac{\text{energy transferred to next level}}{\text{total energy in}} \times 100$$

Question 1: Apply (2 marks)

Calculate the efficiency of biomass transfer from primary consumers and decomposers to secondary consumers.

Question 2: Apply (1 mark)

Identify the producers.

Question 3: Analyse (3 marks)

Organise these organisms into three groups that represent the same trophic level: echidna, lizard, emu, dingo.

Question 4: Analyse (1 mark)

Sequence the appropriate organisms to show an example of the longest food chain.

Question 5: Interpret (3 marks)

Compare the energy available to the eagle and to the cricket. Refer to both Figure 3.5 and Figure 3.6 in your response.

UNIT 4
HEREDITY AND CONTINUITY OF LIFE

4

Chapter 4
Topic 1 DNA, genes and the continuity of life

Topic summary

Deoxyribonucleic acid (DNA) is the molecule that contains all the information required to build and maintain an organism. All known cellular-reproducing organisms have their instructions in the form of DNA.

In prokaryotic organisms, DNA is found freely in the cytoplasm, usually as a single loop. Many prokaryotes also have small, circular rings of DNA called plasmids. In eukaryotic organisms, most of the DNA is found in pieces called chromosomes that are contained within the nucleus. The information required for development and survival is organised along each chromosome in sections called genes. The structures and functions of each organism are controlled by the form and expression of its genes. DNA is also found in the mitochondria and chloroplasts of eukaryotic cells.

DNA is copied and transferred to offspring through the processes of cellular replication, then reproduction. Organisms that reproduce asexually produce genetically identical (to each other and the parent) offspring, whereas those that reproduce sexually produce genetically varied offspring. The combination of genes inherited and environmental interactions affects the expression and function of genes and therefore the survival of every organism.

Biotechnology uses cellular processes to develop useful techniques and products, such as sequencing DNA and recombinant DNA.

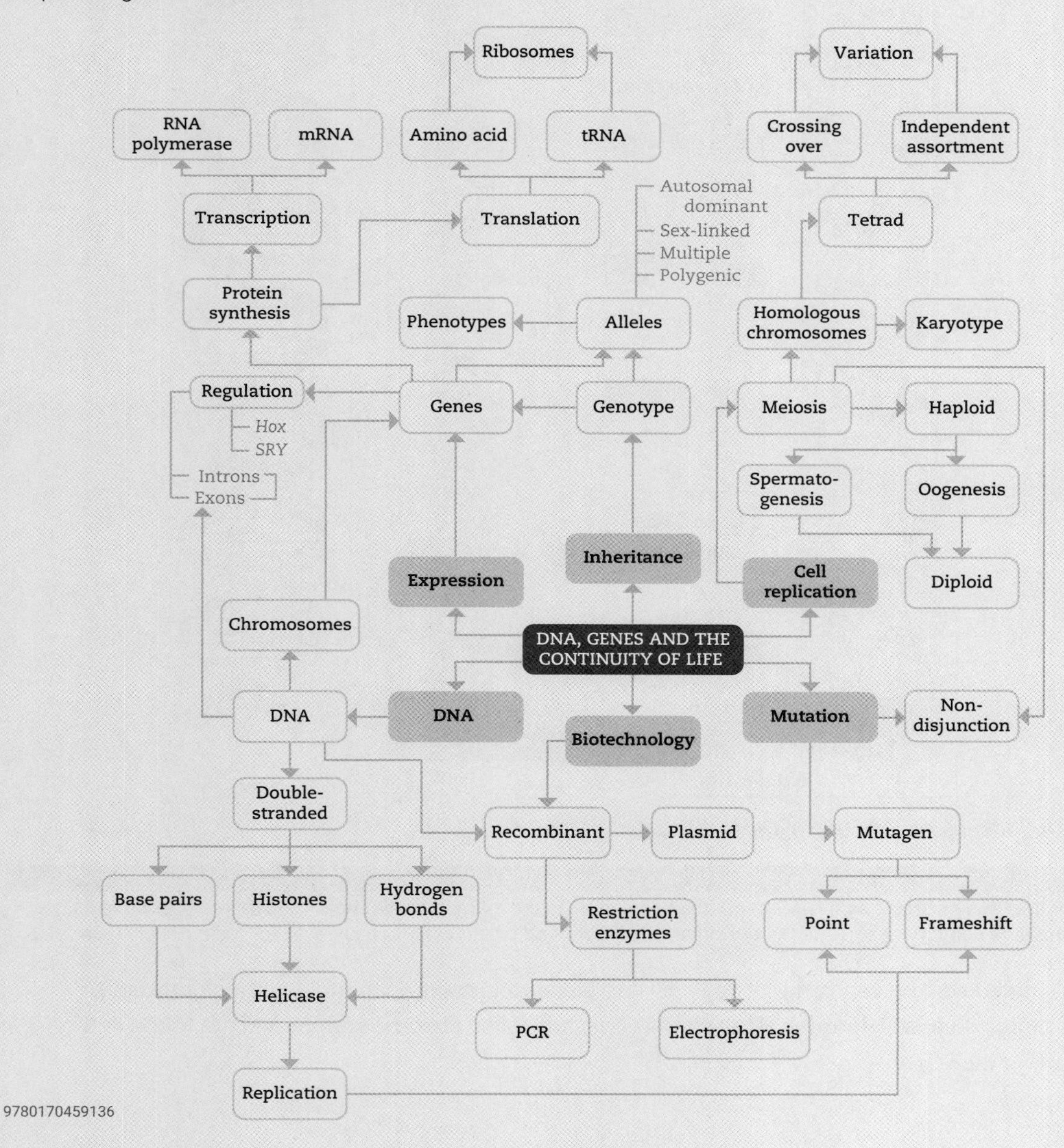

4.1 DNA structure and replication

4.1.1 DNA structure

DNA is a double-stranded nucleic acid that contains the code for the molecules required for life. The **monomer** for DNA is a **nucleotide**. Each nucleotide is made up of a sugar (deoxyribose), a phosphate group and a base (adenine (A), thymine (T), guanine (G) or cytosine (C)) (Figure 4.1).

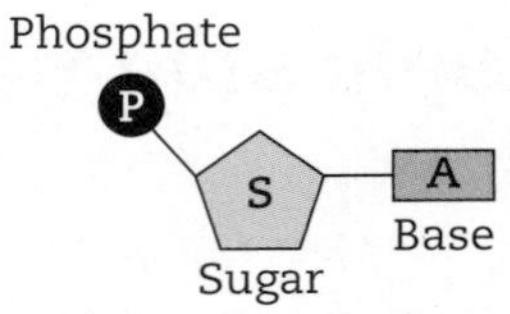

FIGURE 4.1 A nucleotide: typically, the sugar is represented by a pentagon, the phosphate by a circle and the base by another shape, often with the capital letter of the base inside the shape.

The nucleotides are joined together along the sugar–phosphate backbone and in a specific base sequence. It is these sequences that form genes. One strand contains the code (sense strand or coding strand) that is read in a 5' to 3' direction. The second strand is formed as nucleotides are matched through complementary base-pairing with the (antisense strand or template strand). Adenine pairs with thymine. Guanine pairs with cytosine. Each base pair is held together by hydrogen bonds, usually represented by dotted or dashed lines between the bases of each strand. The double-stranded molecule is coiled into an alpha (α) helix, also known as a double helix (Figure 4.2).

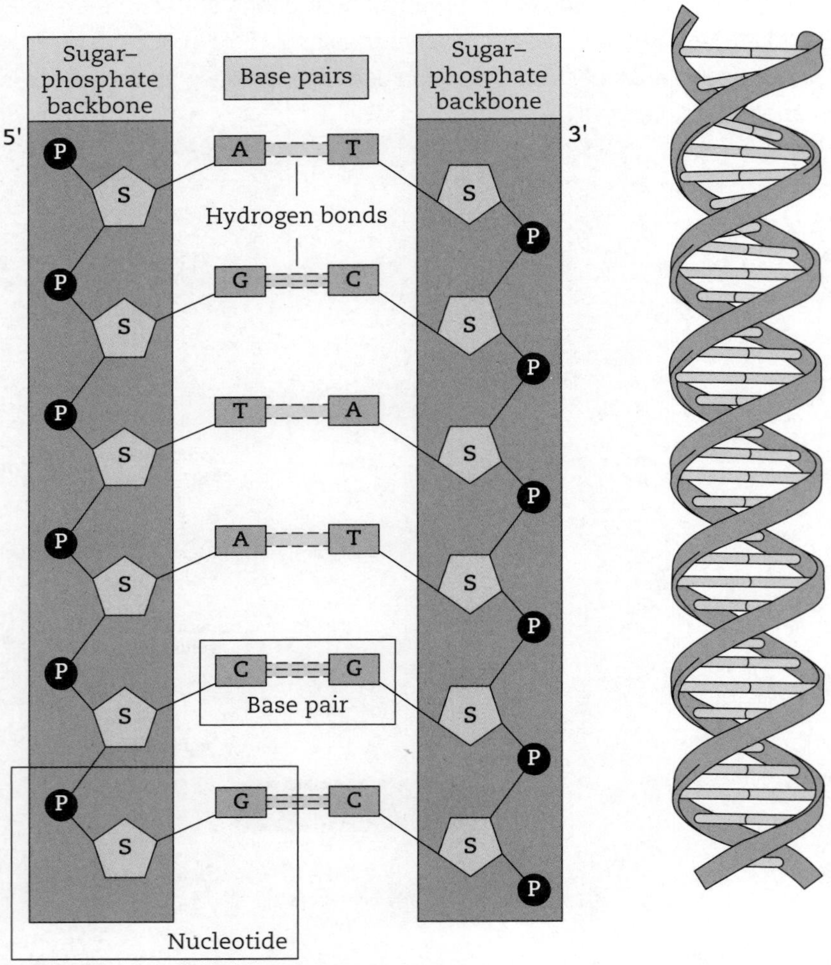

FIGURE 4.2 The structure of a DNA molecule

Hint

Think of the DNA molecule as a ladder, with the base pairs as the rungs and the sugar–phosphate backbone as the sides. Imagine holding the top and the bottom of the ladder and twisting it into a spiral to form the double helix.

DNA molecules are very long. To fit them into eukaryotic cells, each molecule is tightly coiled around proteins called **histones**. These structures are called **chromosomes** and are found in the nucleus (Figure 4.3).

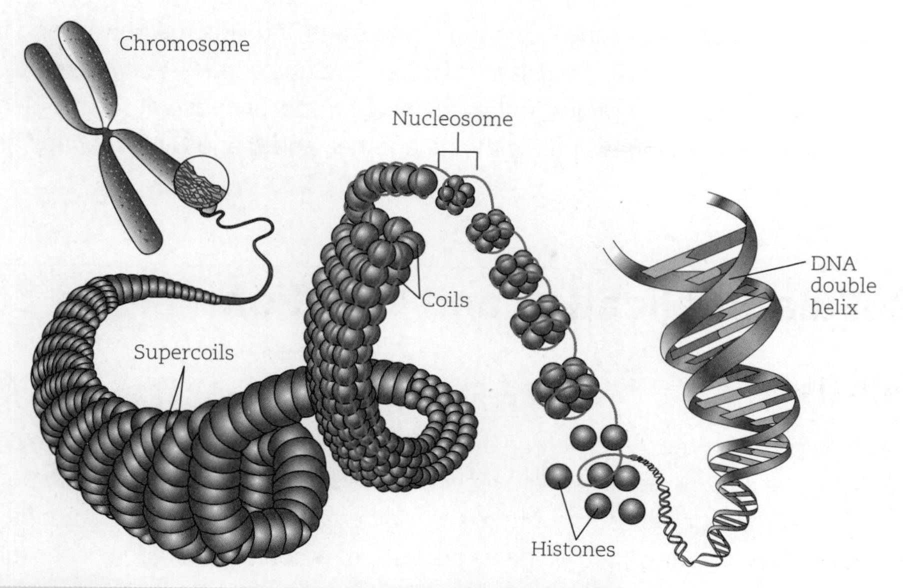

FIGURE 4.3 The DNA double helix coiled around histones to form a chromosome

In prokaryotic cells, DNA occurs in a circular structure, without histones and in the cytosol. In eukaryotic cells, DNA is also found in mitochondria and chloroplasts.

4.1.2 DNA replication

Each new cell that is produced requires its own copy of the DNA molecule(s). In eukaryotic cells, DNA replication is the process that produces new copies of the chromosomes. This is summarised in Figure 4.4.

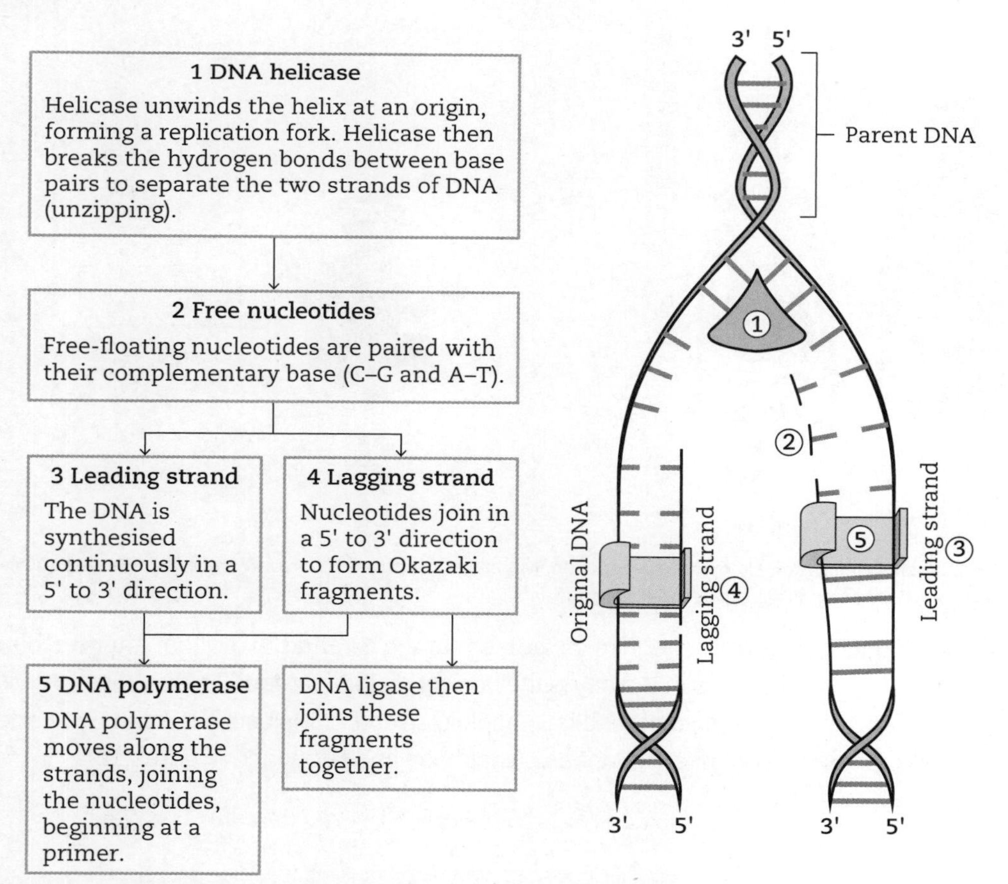

FIGURE 4.4 The steps of DNA replication, including the roles of helicase and DNA polymerase

As DNA polymerase moves down the unwound DNA strand, it builds the new strand from free-floating nucleotides surrounding the existing strand. The nucleotides that make up the new strand are paired with partner nucleotides in the template strand; because of their molecular structures, A and T nucleotides always pair with one another, and C and G nucleotides always pair with one another.

4.2 Cellular replication and variation

4.2.1 Meiosis

Humans have 23 pairs of chromosomes, giving a **diploid** ($2n$) number of 46 (2×23). The diploid number results from the **zygote** (fertilised egg) receiving one set of **paternal** chromosomes and one set of **maternal** chromosomes. Chromosome 1 from the mother and chromosome 1 from the father are called **homologous** chromosomes (homologues) because they contain the same genes (although not necessarily the same forms of the genes – section 4.1).

Hint
Different forms of genes are called alleles.

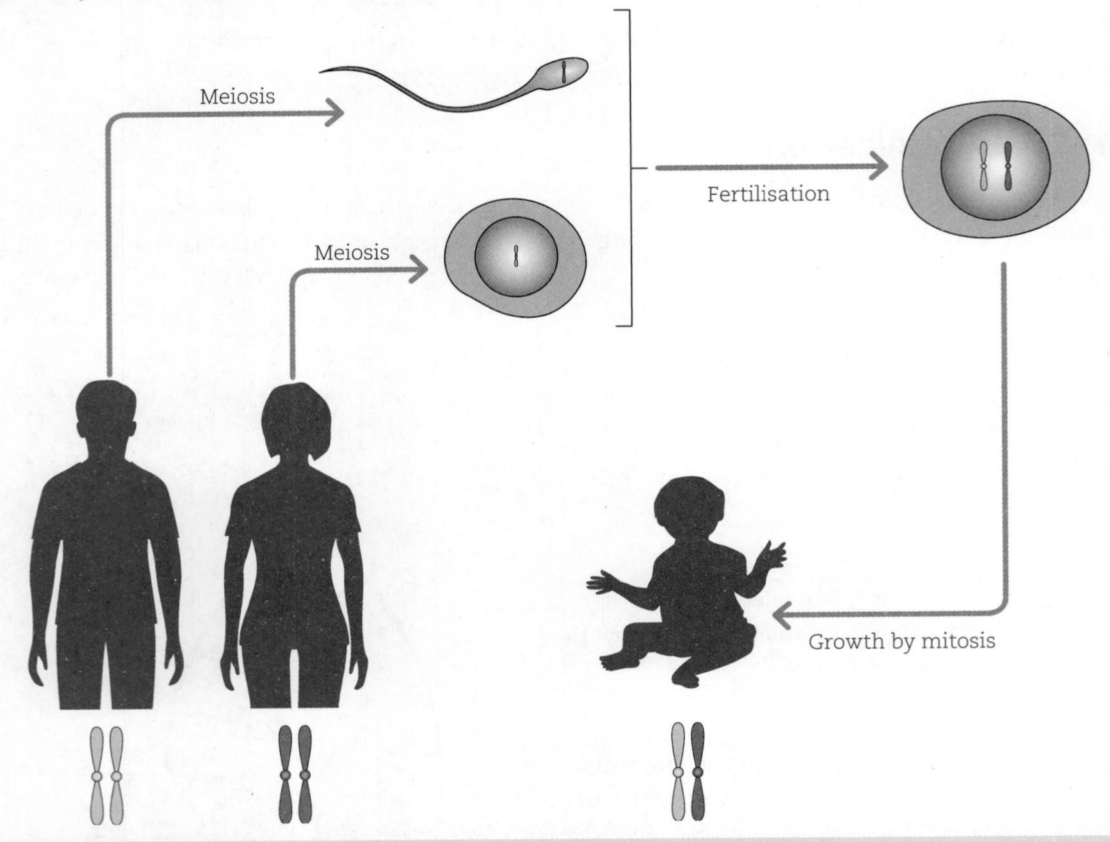

FIGURE 4.5 Meiosis is important for gamete production. Mitosis is important for growth and repair. Humans have 23 pairs of chromosomes. Only one pair is shown for simplicity.

Sexually reproducing organisms require **gametes** that are **haploid** (n) so that after **fertilisation** the total number of chromosomes in the cell is the correct diploid number. This requires a cellular replication process called meiosis, that results in haploid sex cells. Meiosis occurs in two parts, each consisting of four phases: prophase, metaphase, anaphase and telophase (Figure 4.6).

Meiosis I: Reduction division – to separate homologous chromosomes

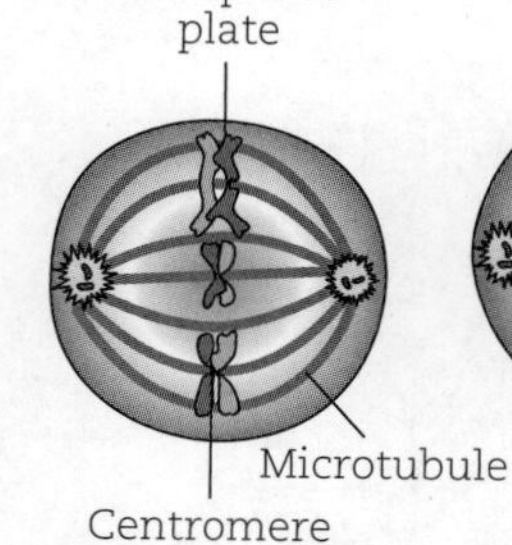
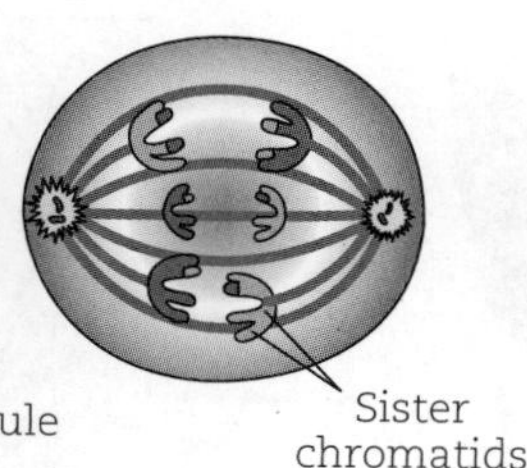
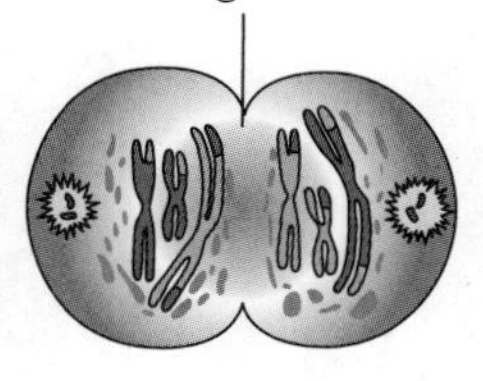

Meiosis II: Equational division – to separate chromatids

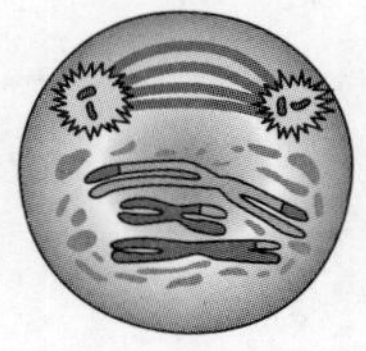
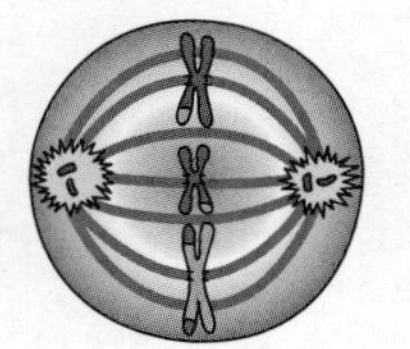
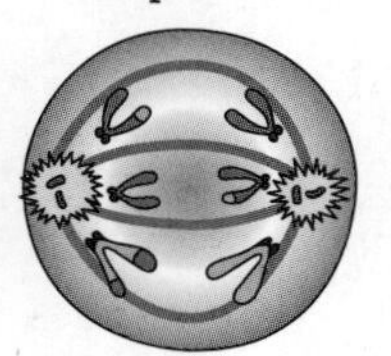
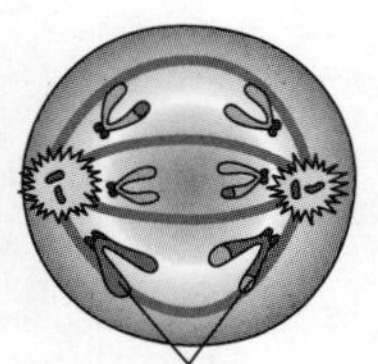
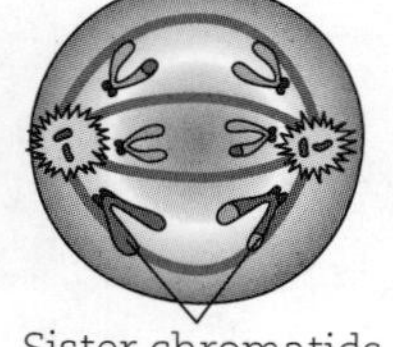
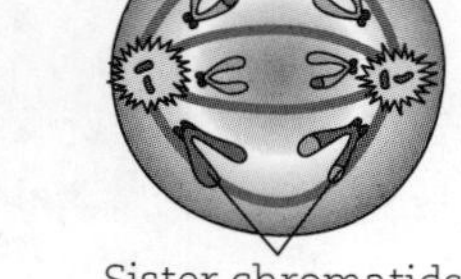

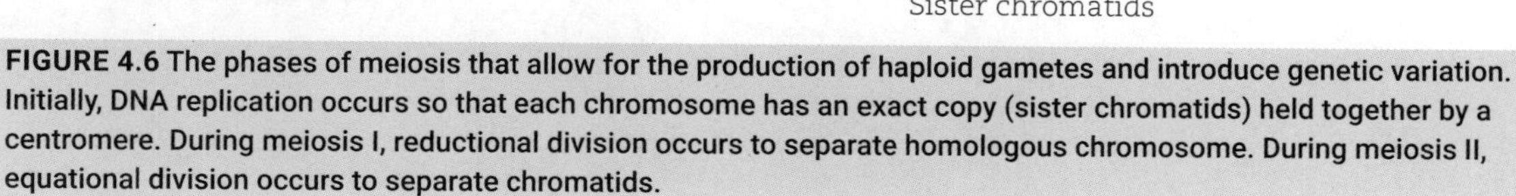
FIGURE 4.6 The phases of meiosis that allow for the production of haploid gametes and introduce genetic variation. Initially, DNA replication occurs so that each chromosome has an exact copy (sister chromatids) held together by a centromere. During meiosis I, reductional division occurs to separate homologous chromosome. During meiosis II, equational division occurs to separate chromatids.

> **Hint**
> The cells formed at the end of meiosis I are considered haploid because they have just one chromosome from each homologue pair even though the chromosomes still consist of two sister chromatids.

> **Hint**
> This second division without replication is how the daughter cells end up with only one copy of each chromosome.

> **Hint**
> Meiosis II looks a lot like mitosis, but it results in four genetically different haploid cells.

> **Hint**
> The basic movement of chromosomes is the same for each phase in the meiotic divisions. What differs is the number of chromatids moving as a group.

4.2.2 Variation

Variation is important because it means that the offspring of sexually reproducing organisms are genetically different from each other and their parents.

Crossing over

Crossing over occurs during prophase I. It means that maternal and paternal sections of DNA (genes) are exchanged between non-sister **chromatids**. Genes occur in a new or different combination from how they exist in the organism's body cells.

> **Hint**
> The importance of variation is discussed when we look at evolution.

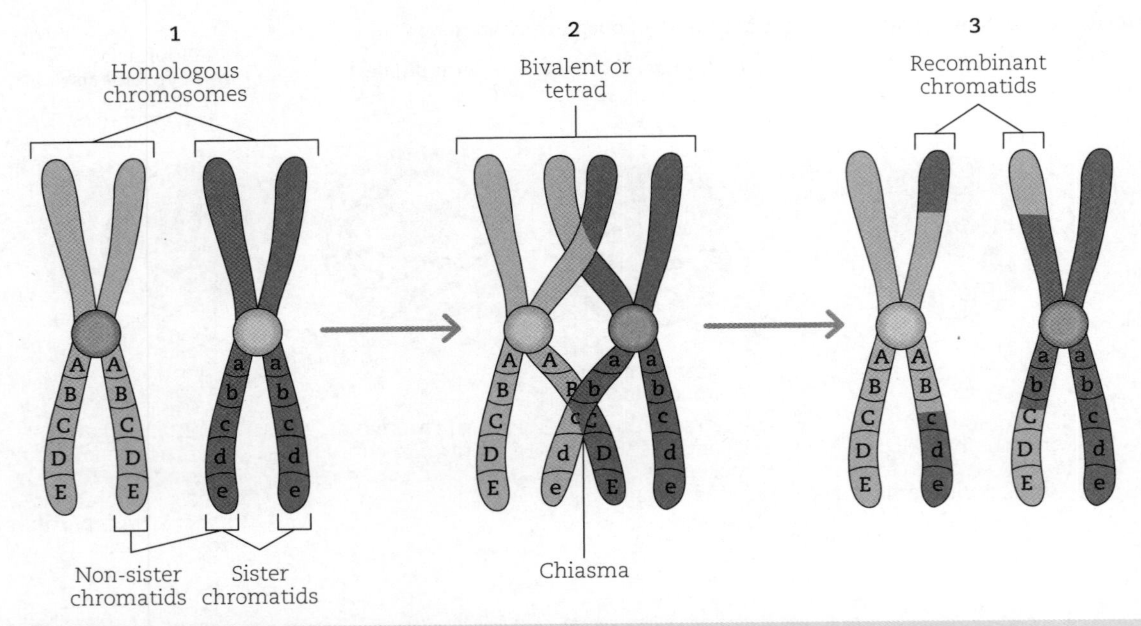

FIGURE 4.7 Variation is increased by the crossing over of sister chromatids during meiosis I. It results in new combinations of gene forms on a chromosome.

Independent assortment

When the chromatid groups align at metaphase, they do so independently of every other chromatid group. There are two alignment possibilities or 2^n, where *n* is the number of chromosomes per set or haploid number.

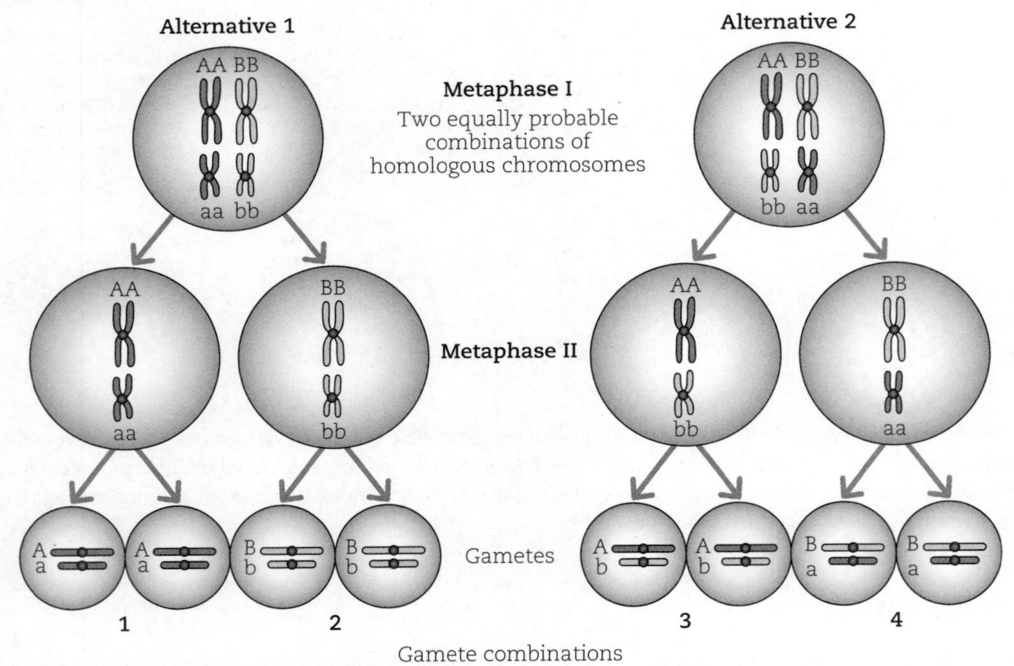

FIGURE 4.8 Sister chromatids align independently of every other chromatid pair during metaphase I and metaphase II. This diagram shows the possible outcomes for an organism, where $n = 2$ and the diploid number is 4 (in humans, $n = 23$ and the diploid number is 46).

Resulting gametes have different combinations of chromosomes, resulting in variation.

It is highly unlikely that any two haploid cells resulting from crossing over and independent assortment will have the same combination of gene forms.

Random fertilisation

Males produce 1500 sperm per second and an average of 100 million sperm are released at one time. Each sperm is genetically different from the others produced by that male. Only *one* of these sperm can fertilise an egg.

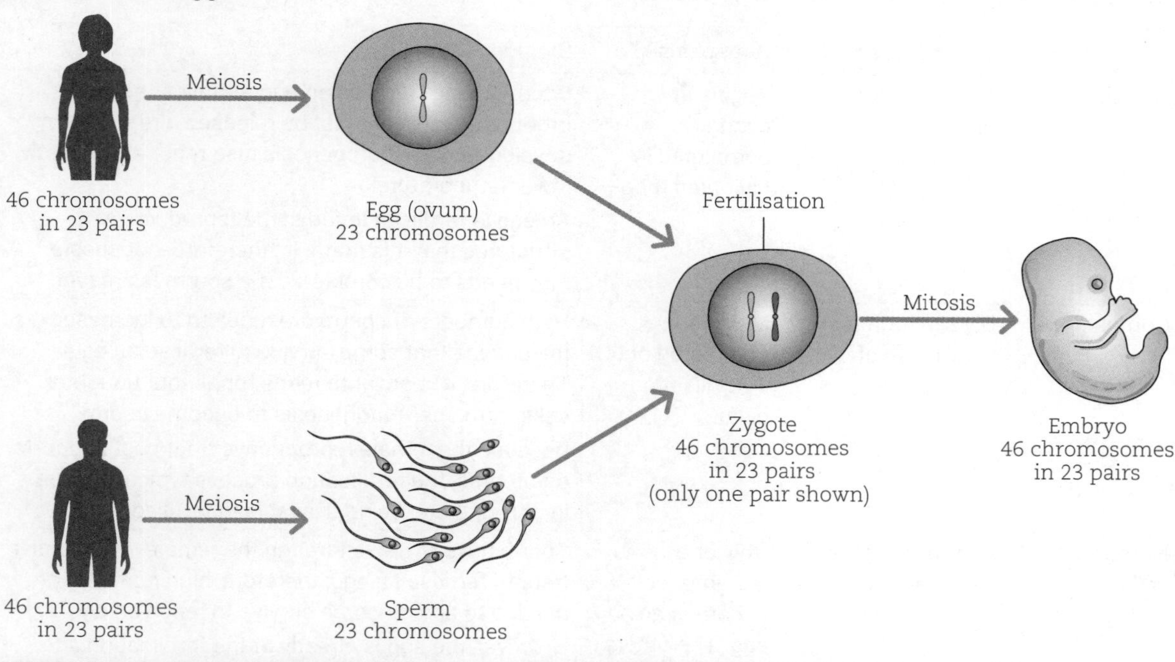

FIGURE 4.9 The random fertilisation of an egg by one of millions of sperm.

Hint

A sperm cell, with 2^n possible chromosome combinations, fertilises an egg cell, which also has 2^n possible chromosome combinations. That is $2^{23} \times 2^{23}$ unique combinations in humans. This number does not include the changing combinations due to crossing over!

Hint

The exception is identical twins, which form from one zygote and are genetically identical.

4.2.3 Oogenesis versus spermatogenesis

Although meiosis occurs in both males and females, the rate and timing of this cell replication is different.

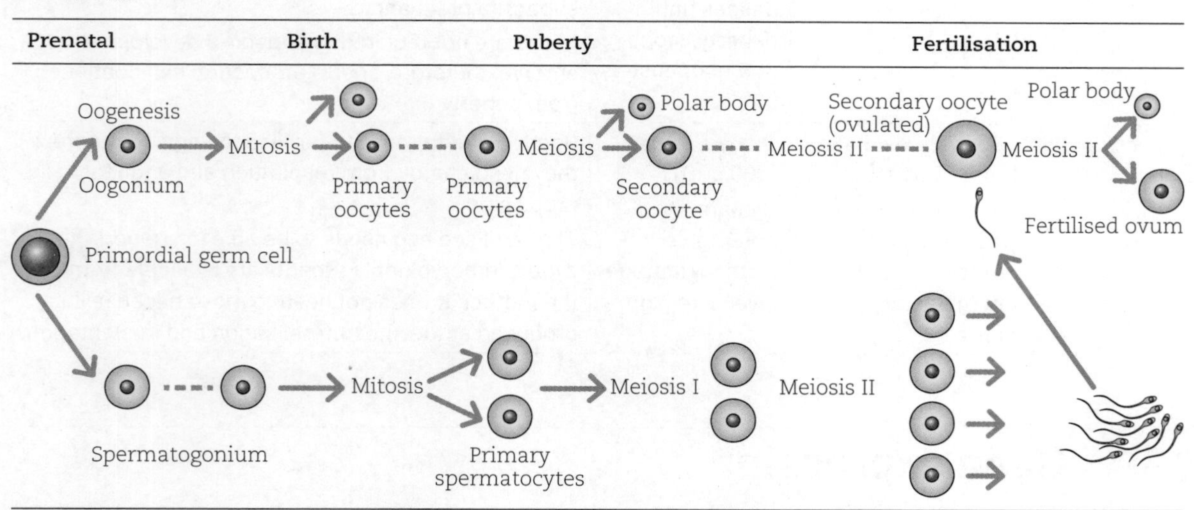

FIGURE 4.10 Comparing and contrasting the processes of spermatogenesis and oogenesis

The similarities in spermatogenesis and oogenesis are that they both:

- result in haploid gametes that are genetically variable
- occur in gonads
- are processes controlled by hormones.

These are important because for sexual reproduction to be successful, both males and females need to produce haploid gametes that result in a diploid zygote after fertilisation.

The differences between spermatogenesis and oogenesis are summarised in Table 4.1.

TABLE 4.1 Differences in spermatogenesis and oogenesis

	Spermatogenesis	Oogenesis	Significance
Location	All completed in testes	Begins in ovaries, completed in Fallopian tube	Sperm cannot fertilise an egg at their site of production so they must be released already developed. Spermatogenesis also requires a slightly lower temperature. An egg is fertilised in the same reproductive structures that it is made in; therefore, oogenesis only needs to be completed if a sperm is present.
Completion	Division completed regardless of fertilisation	Meiosis II completed only if fertilisation occurs	High numbers of sperm are required to increase the chance that some survive to fertilise an egg. Therefore, it is advantageous for all four daughter cells from spermatogenesis to become sperm. Because the female reproductive tract best supports one foetus, the fertilisation process typically results in only one egg being available to fertilise.
Number of cells	All four daughter cells become sperm.	Only one daughter cell becomes an egg. The others become polar bodies.	Sperm have to move through the female reproductive tract to fertilise an egg; therefore, high numbers are needed to ensure some survive to reach an egg. However, the egg is already at the fertilisation site in the female reproductive tract, so there is no requirement for large numbers of eggs to survive a journey.
Timing	Is a continuous process from the onset of puberty.	Begins during foetal development; pauses until puberty; stops at menopause.	The whole body needs to be ready to support a fertilised egg; therefore, meiosis pauses until this is the case. It stops again once the body is less likely to support a pregnancy. Males are not required to support a developing embryo; therefore, sperm production can continue from puberty.
Specialisation	Sperm are smaller (4.5–5.0 μm) Sperm are motile. Tail and high mitochondrial numbers are present.	Ova are larger (100 μm) Ova are non-motile. Nutrient storage levels are high.	Sperm need to move to fertilise the egg; therefore, they need energy from respiration and a tail for propulsion. The fertilised egg needs to be able to support the zygote/embryo until nutrients are available from the mother. It does not need to move because it is produced at the site of fertilisation and implantation.

4.3 Gene expression

4.3.1 Genes

Genes are the molecular units of heredity; they are region(s) of DNA that are made up of sequences of nucleotides. A **genome** is the complete set of genetic material in the chromosomes of an organism, including its genes and DNA sequences.

Genes are found in fixed positions on chromosomes. They are small sections of DNA that code for proteins or RNA and contain the instructions for individual **characteristics** or **traits**. Genes exist in different forms called **alleles** and are passed from parents to offspring. The sequences of non-coding DNA between genes are called intergenic regions.

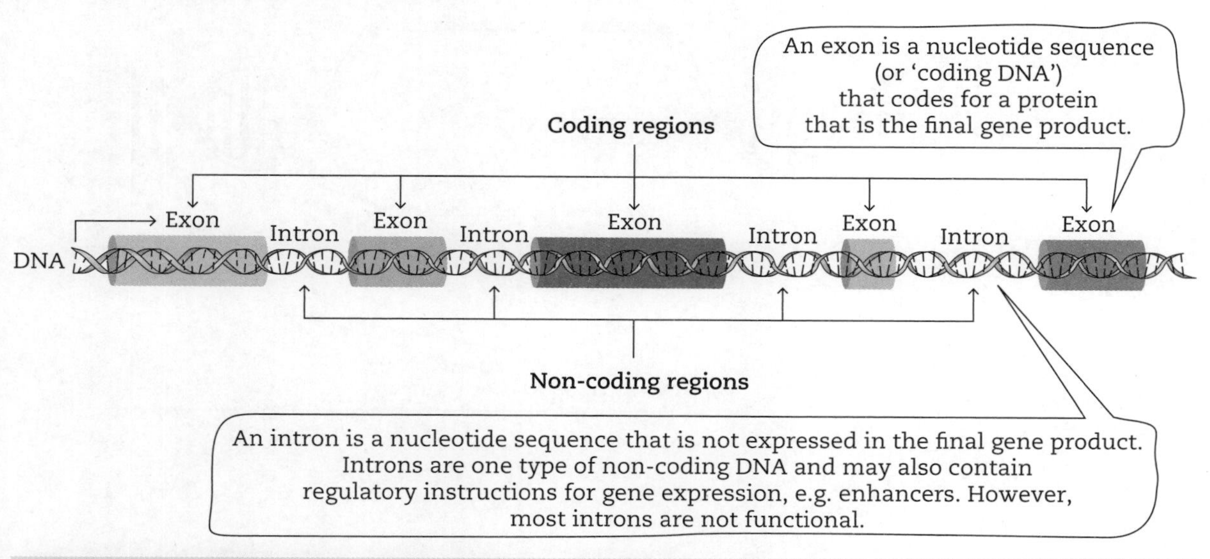

FIGURE 4.11 The difference between introns and exons within a gene

Hint
INtrons Intervene.
EXons are Expressed.

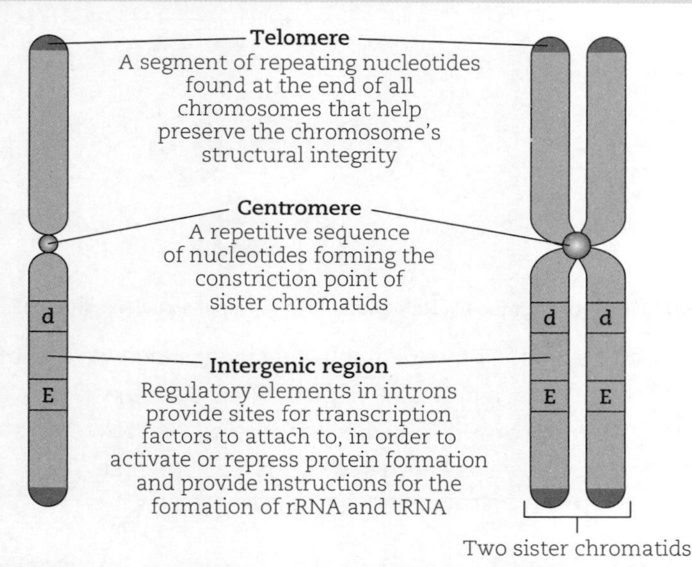

FIGURE 4.12 Genes only make up 1–1.5% of DNA. The remainder is non-coding DNA with sequences that act as regulatory elements, determining when and where genes are turned on and off, as well as instructions for the formation of rRNA and tRNA.

The location and role of regulatory elements and other functions of non-coding DNA is not yet completely understood.

4.3.2 Gene expression and protein synthesis

Gene expression is the process by which the information encoded in the sequence of nucleotide bases in a gene is used to direct the assembly of a protein via protein synthesis. The sequence of nucleotide bases in the gene is read in groups of three (**codons**). Each codon corresponds to a specific amino acid of the 20 types used to build a protein.

Protein synthesis consists of two stages: transcription and translation (Figure 4.13, page 94). Transcription occurs in the nucleus and results in a strand of messenger RNA (mRNA). Translation occurs in the cytoplasm and results in a polypeptide.

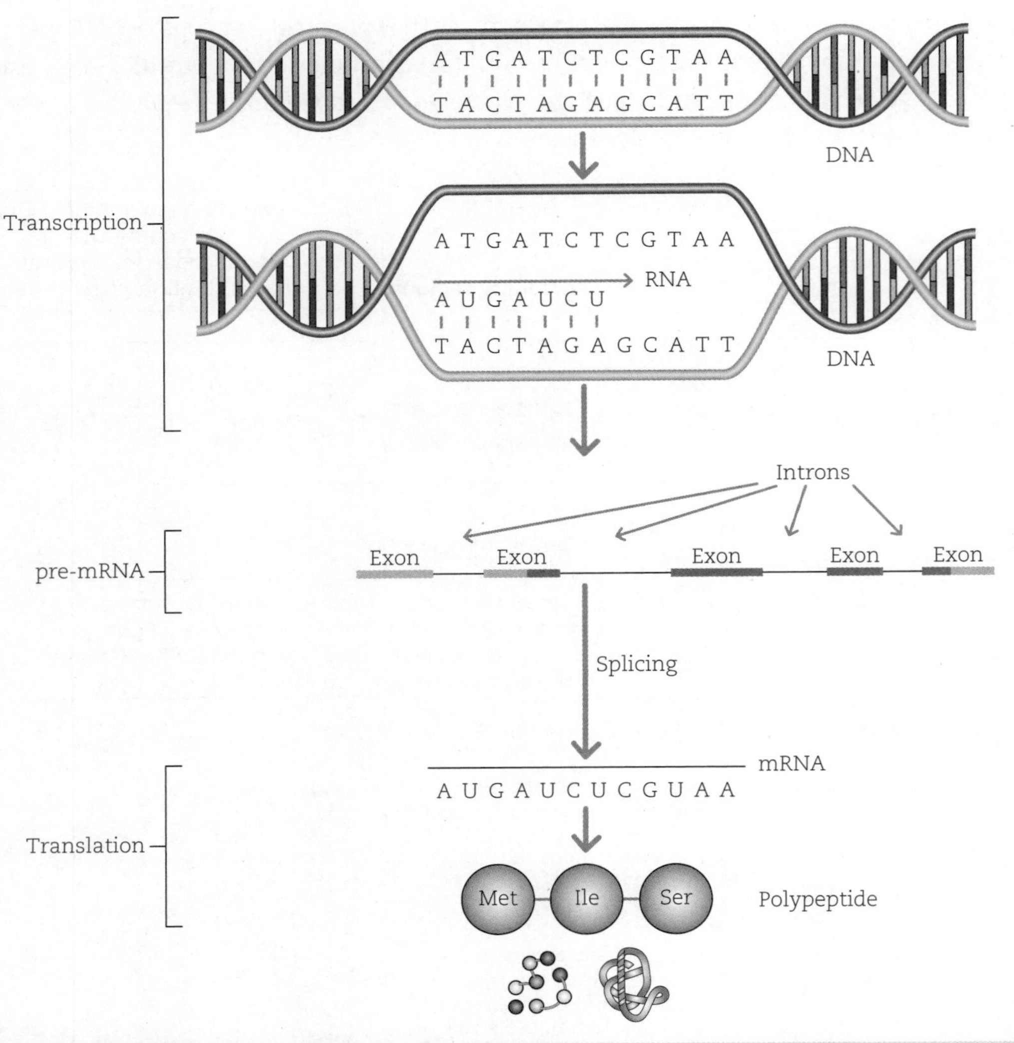

FIGURE 4.13 the steps involved in protein synthesis. Note that mRNA leaves the nucleus after splicing.

Transcription produces a copy of the required nucleotide base sequence (gene) in the DNA (Figure 4.14). During protein synthesis, a molecule of pre-mRNA is synthesised in a 5' to 3' direction, from RNA nucleotide bases that are **complementary** to those in the DNA template, although in RNA uracil (U) replaces thymine. This process is **catalysed** by RNA polymerase.

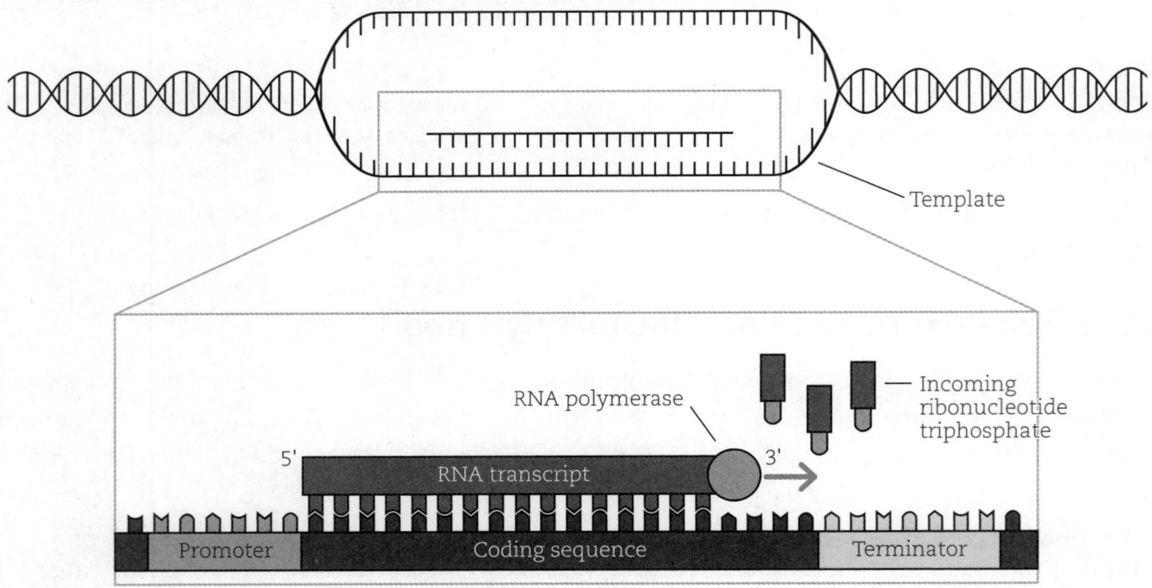

FIGURE 4.14 Transcription: copying DNA to synthesise mRNA

Several types of RNA can be encoded by a DNA strand, including rRNA and tRNA. After splicing has occurred to remove any **introns**, the **exons** are joined together and the mature mRNA can leave the nucleus.

One pre-mRNA molecule can be spliced in many ways, allowing one gene to produce multiple proteins.

Translation occurs in the cytoplasm and produces a sequence of amino acids (a **polypeptide**) determined by the order of nucleotide bases in the mRNA. Each amino acid is represented by at least one codon. The start codon is AUG, which is the amino acid methionine. This codon sets the reading frame, determining where the grouping of bases into triplets starts.

In the nucleotide sequence below, the box shows where the start codon is. The bases are 'read' in groups of three from this point until a stop codon is reached.

A G G C [A U G] U U U C G G A A A G U G . . .

Codons: AUG UUU CGG AAA GUG . . .

The triplet code and associated amino acids are listed in Figure 4.15.

You can see that the five codons above are translated as: methionine, phenylalanine, arginine, lysine and valine.

Ala	alanine
Arg	arginine
Asn	asparagine
Asp	aspartic acid
Cys	cysteine
Gln	glutamine
Glu	glutamic acid
Gly	glycine
His	histidine
Ile	isoleucine
Leu	leucine
Lys	lysine
Met	methionine
Phe	phenylalanine
Pro	proline
Ser	serine
Thr	threonine
Trp	tryptophan
Tyr	tyrosine
Val	valine

First base	Second base: U	Second base: C	Second base: A	Second base: G	Third base
U	UUU, UUC – Phe UUA, UUG – Leu	UCU, UCC, UCA, UCG – Ser	UAU, UAC – Tyr UAA – Stop UAG – Stop	UGU, UGC – Cys UGA – Stop UGG – Trp	U C A G
C	CUU, CUC, CUA, CUG – Leu	CCU, CCC, CCA, CCG – Pro	CAU, CAC – His CAA, CAG – Gln	CGU, CGC, CGA, CGG – Arg	U C A G
A	AUU, AUC, AUA – Ile AUG – Met/Start	ACU, ACC, ACA, ACG – Thr	AAU, AAC – Asn AAA, AAG – Lys	AGU, AGC – Ser AGA, AGG – Arg	U C A G
G	GUU, GUC, GUA, GUG – Val	GCU, GCC, GCA, GCG – Ala	GAU, GAC – Asp GAA, GAG – Glu	GGU, GGC, GGA, GGG – Gly	U C A G

FIGURE 4.15 mRNA codon to amino acid table

Ribosomes (rRNA) assemble around the strand of mRNA, where they cause transfer RNA (tRNA) molecules to bring the correct amino acids according to the triplet code of the base sequence by matching the tRNA **anticodon** to the mRNA codon. The assembled amino acids are joined by peptide bonds as the ribosome moves along the mRNA molecule until a stop codon is reached. This is shown in Figure 4.16 (page 96).

Protein synthesis occurs in all living cells, although there are differences between prokaryotic and eukaryotic cells in ribosome structure and regulatory factors.

Hint
Remember that proteins can have primary, secondary, tertiary and quaternary levels of structure. The amino acid sequence is only the primary level.

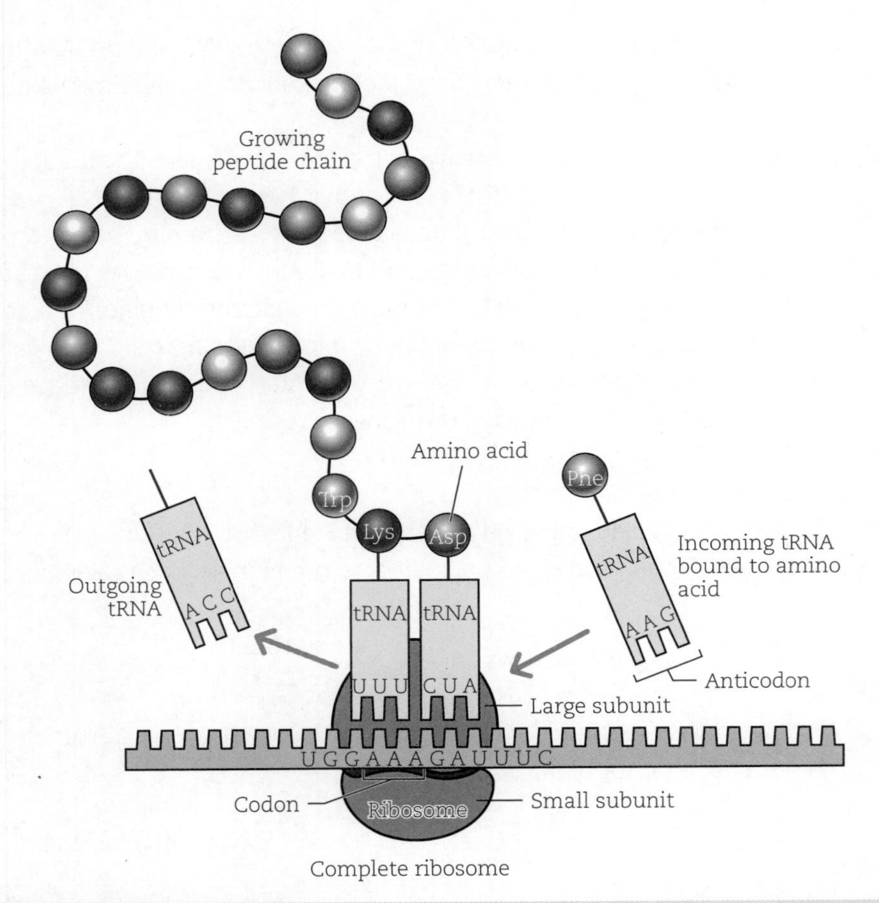

FIGURE 4.16 Translation of mRNA into a polypeptide

4.3.3 Regulating gene expression

Each cell of a multicellular organism carries the same DNA. However, each cell may use this information in different ways, depending on the cell's role. The genes that are expressed are regulated or controlled by proteins or aspects of the cell's environment. The expression of different genes in cells is what produces specialised tissues. The control of how, when and for how long genes are expressed allows for conserved space within the cell (only a section of DNA unwinds from its tightly coiled structure) and energy efficiency (genes are only 'turned on' when they are required).

This regulation can occur in several ways at each step of protein synthesis, but occurs most often at the transcriptional level.

Transcriptional regulation alters the rate of transcription to change gene expression levels, using transcription factors (proteins that assist in the binding of RNA polymerase) and regulatory proteins (e.g. activators, enhancers or repressors).

Translational regulation controls the amount of proteins produced from mRNA. There are also regulatory mechanisms that edit proteins that have already been made.

Ultimately, any processing or modification step during protein synthesis can be regulated. Gene expression can be further affected by intracellular factors that, in turn, reflect environmental factors, including hormones, disease and the activity of other genes.

Epigenetic factors are external modifications (behaviour and/or environment) that change how genes are read by the cell but do not change the DNA sequence. Studies of identical twins are designed to measure what proportion of a given trait results from the genes present compared to environmental influences. Identical twins are monozygotic because they come from a single fertilised egg that splits in two. These twins share 100% of their genes, which provides researchers with an opportunity to study the effect of environment on gene expression.

Hox genes

The *Hox* genes (homeobox genes) are a set of transcription factor genes that specify the body plan in an embryo and produce the body pattern of head, thorax and abdomen, as well as cell differentiation. *Hox* proteins are involved in several steps of gene expression. The genes are highly conserved between species (see Figure 4.17) because every organism requires the development of body structure and differentiated cells. In mammals, *Hox* genes are found on four chromosomes. These genes provide evidence of common ancestry (see Unit 4 Topic 2).

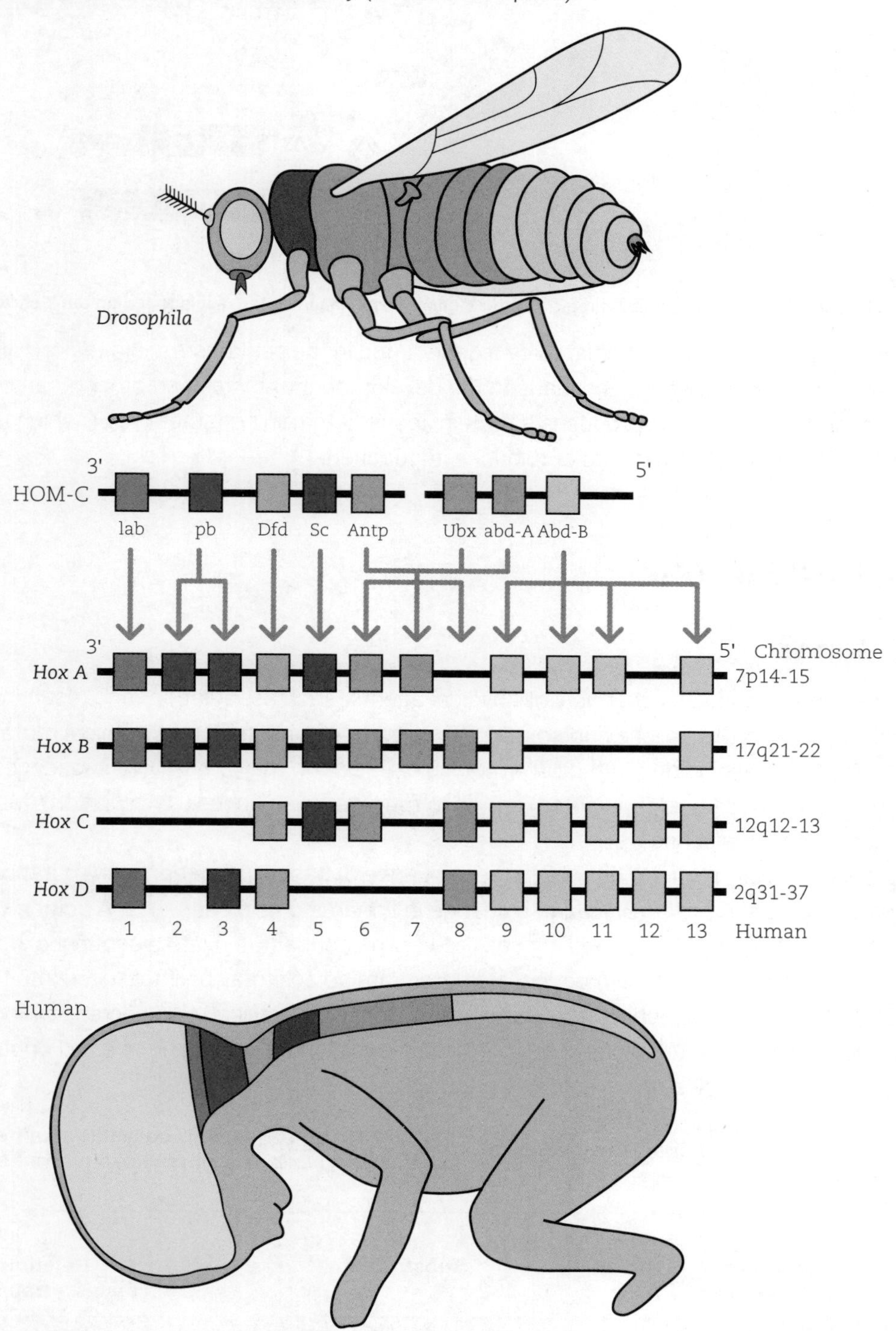

FIGURE 4.17 The similarity in *Hox* genes for humans and fruit flies (*Drosophila*)

SRY gene

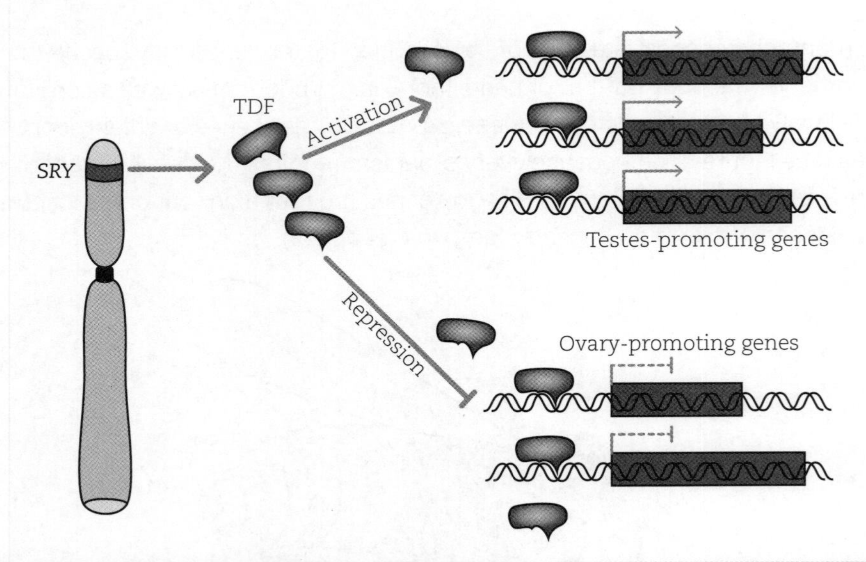

FIGURE 4.18 The *SRY* or sex-determining region of the Y gene. It is Y linked because it is found on the Y chromosome.

Male embryonic development is initiated by a gene found in the sex-determining region (*SRY*) on the short arm of the Y chromosome and leads to the development of testes and other male phenotypes. The DNA encoded by the gene produces a protein, testis-determining factor (TDF), which is a transcription factor that can activate or enhance its function.

4.4 Mutations

4.4.1 Genes

> **Hint**
> Think of this like a word:
> Original word: litter
> Insertion: glitter
> Deletion: liter
> Substitution: bitter

Mutations can be caused by chemical or physical agents (such as radiation) and result in a permanent change to the genetic code of an organism by altering its DNA. Chemical mutagens typically affect the base pairing of the DNA molecule. Radiation may break DNA sequences (ionising radiation) or cause additional (unwanted) bonds to form between bases (UV radiation). Mutations are often associated with diseases, especially cancers.

A mutation typically refers to change in the sequence of bases in a gene. Such mutations can generally only be observed if they lead to a change in the amino acid sequence. A point mutation occurs when a single base is inserted into, deleted from or substituted into a sequence. If these changes occur and alter the mRNA codon, a different amino acid may be introduced into the protein. Alternatively, the polypeptide chain may terminate early. It can also lead to a decrease in production or expression of the gene. A change in the DNA sequence is more damaging to a cell or organism because it affects all copies of the encoded protein.

> **Hint**
> Think of a reading frame like this sentence: 'The ant ate out one eve.'
> Insertion: The ant lat eou ton eev e.
> Deletion: The ant teo uto nee ve.
> Substitution: The ant ate put one eve.

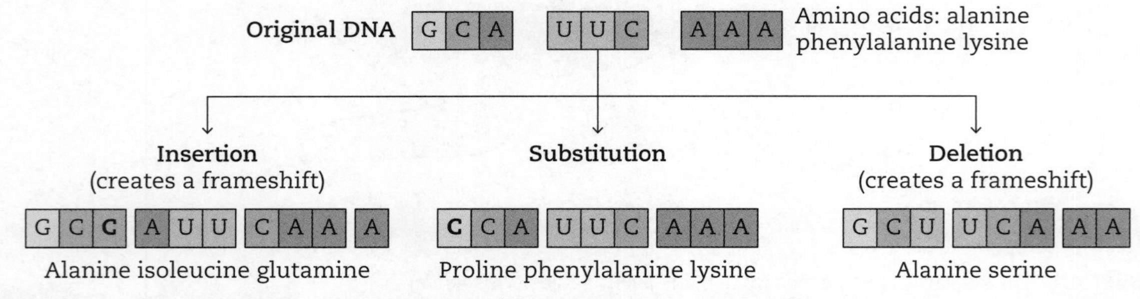

FIGURE 4.19 Summarising the effect of point mutations on an amino acid sequence

A real-life example occurs in the beta haemoglobin chain of 147 amino acids. A single-base mutation changes the primary amino acid sequence, resulting in sickle cell anaemia. This substitution is depicted in Table 4.3.

TABLE 4.3 Single base mutation that results in sickle cell anaemia

Sequence for wild-type haemoglobin												
ATG	GTG	CAC	CTG	ACT	CCT	GAG	GAG	AAG	TCT	GCC	GTT	ACT
Start	Val	His	Leu	Thr	Pro	Glu	Glu	Lys	Ser	Ala	Val	Thr
Sequence for mutant (sickle-cell) haemoglobin												
ATG	GTG	CAC	CTG	ACT	CCT	GTG	GAG	AAG	TCT	GCC	GTT	ACT
Start	Val	His	Leu	Thr	Pro	Val	Glu	Lys	Ser	Ala	Val	Thr

If a gamete containing a mutation is subsequently fertilised, the zygote will have a new version of the gene that has not previously existed. The individual has a version of a gene that is not present in either parent. It also means that all future offspring from the adult who develops from the zygote may inherit the new version of the gene.

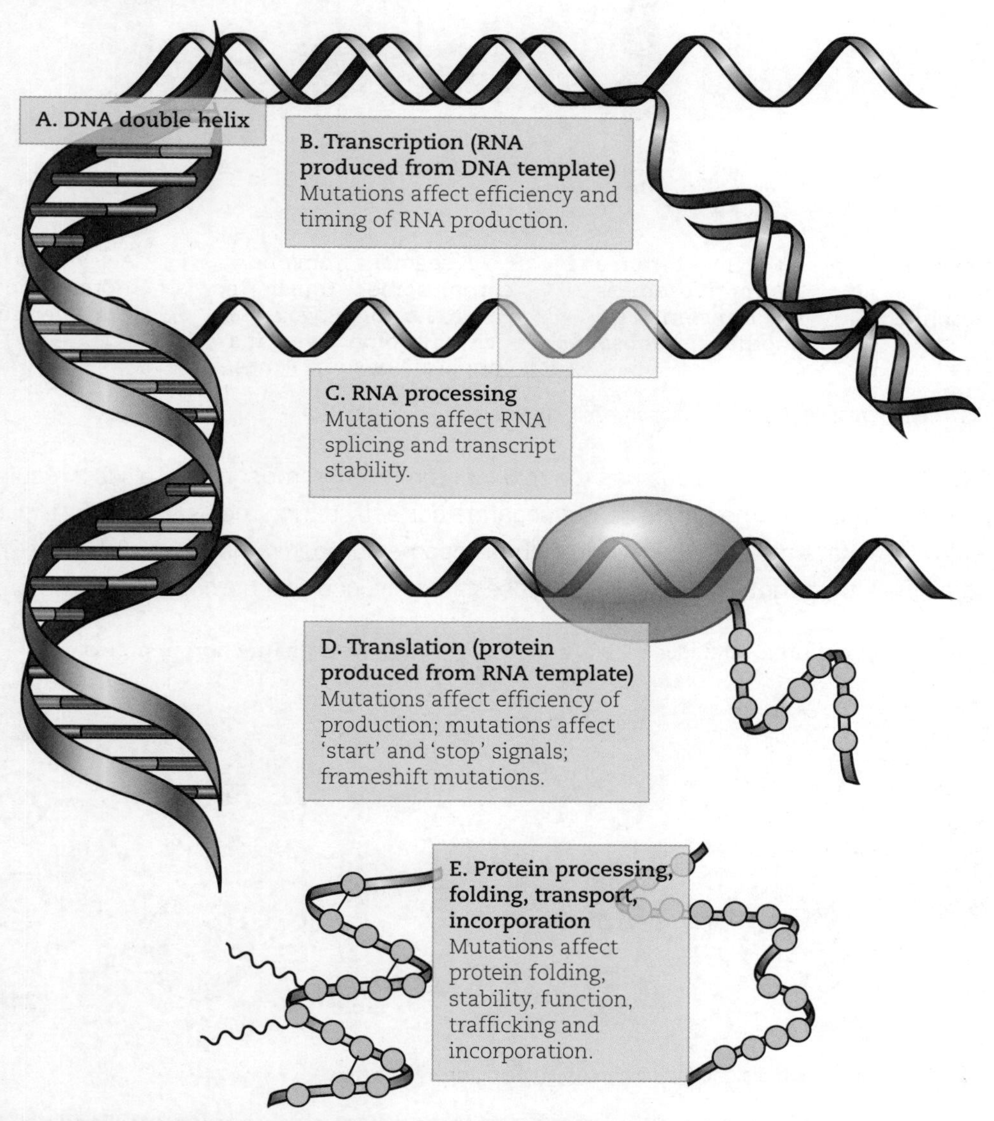

FIGURE 4.20 Stages of protein synthesis that can be affected by mutations

4.4.2 Chromosomes

Large-scale mutations that involve changes in chromosome structure can affect the functioning of many genes, resulting in major changes in phenotype. These mutations include the deletion or insertion of one or more genes, the inversion of genes on a chromosome or the exchange of DNA segments between non-homologous chromosomes (Figure 4.21).

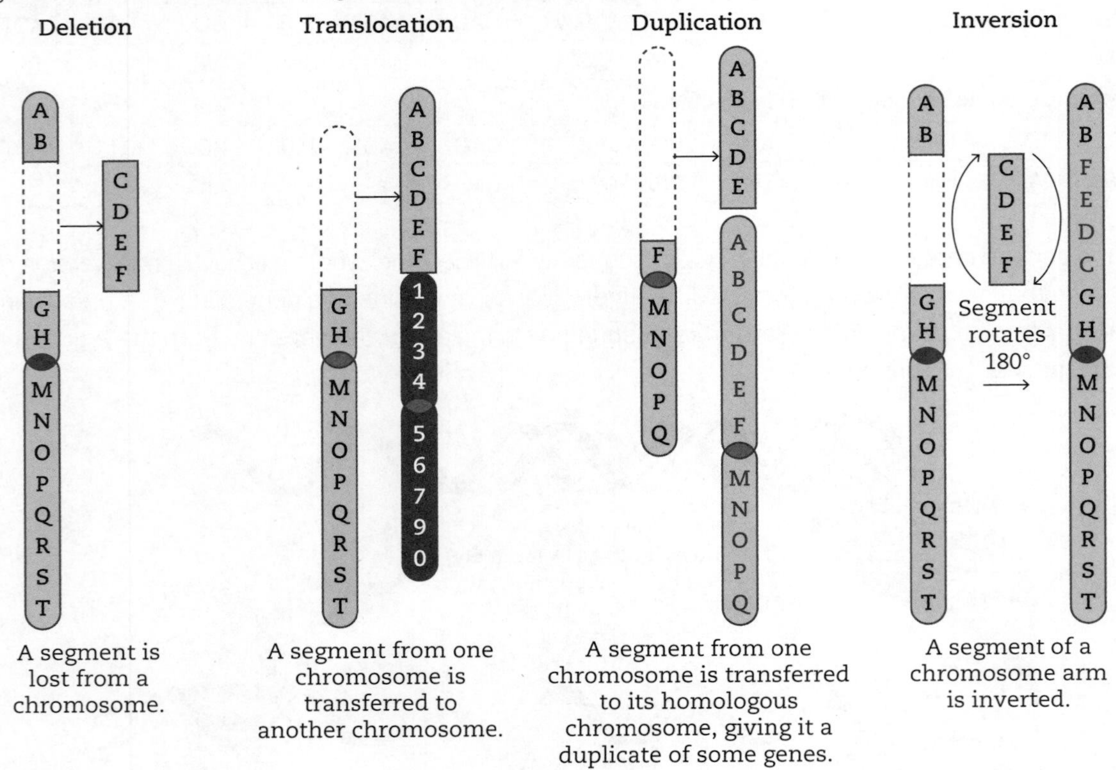

FIGURE 4.21 The four main types of chromosome mutations

Aneuploidy is when a cell nucleus has too many or too few chromosomes. This is a result of **non-disjunction** when chromosomes do not separate correctly during meiosis. After fertilisation, the zygote does not have the correct $2n$ diploid chromosome number; rather, it has $2n + 1$ or $2n - 1$ (Figure 4.22). Often, the zygote is not viable because it has the incorrect number of chromosomes.

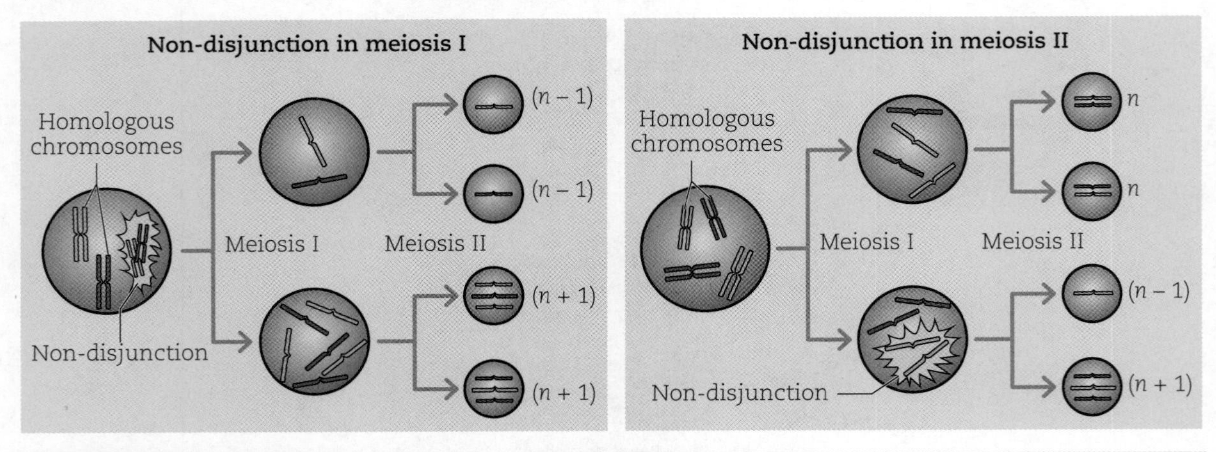

FIGURE 4.22 The effects of non-disjunction on chromosome number in gametes

One example of a viable aneuploidy is trisomy 21 (three copies of chromosome 21), which is more commonly known as Down syndrome.

The syllabus requires you to be able to match the chromosomes in a **karyotype** with the description of a disorder. It is important to recognise how human karyotypes are presented and to be able to identify which chromosome has extra copies *or* is missing copies. Table 4.4 shows the most common aneuploidies in humans.

Hint
You do not have to remember all the syndromes, only be able to identify them from information provided.

TABLE 4.4 Outcome of aneuploidy in humans

Disorder	Inheritance	Abnormality
Down syndrome	Autosomal	47 chromosomes, trisomy 21
Edward's syndrome	Autosomal	47 chromosomes, trisomy 18
Patau syndrome	Autosomal	47 chromosomes, trisomy 13
Klinefelter syndrome	Sex	47 chromosomes, XXY
Turner syndrome	Sex	45 chromosomes, X
Trisomy X (super female)	Sex	47 chromosomes, XXX
Supermale	Sex	47 chromosomes, XYY
Cri du chat syndrome	Autosomal	Partial deletion of chromosome 5
DiGeorge or velocariofacial syndrome	Autosomal	Microdeletion of chromosome 22
Prader-Willi syndrome	Autosomal	Deletion of paternal chromosome 15
Angelman syndrome	Autosomal	Deletion of maternal chromosome 15

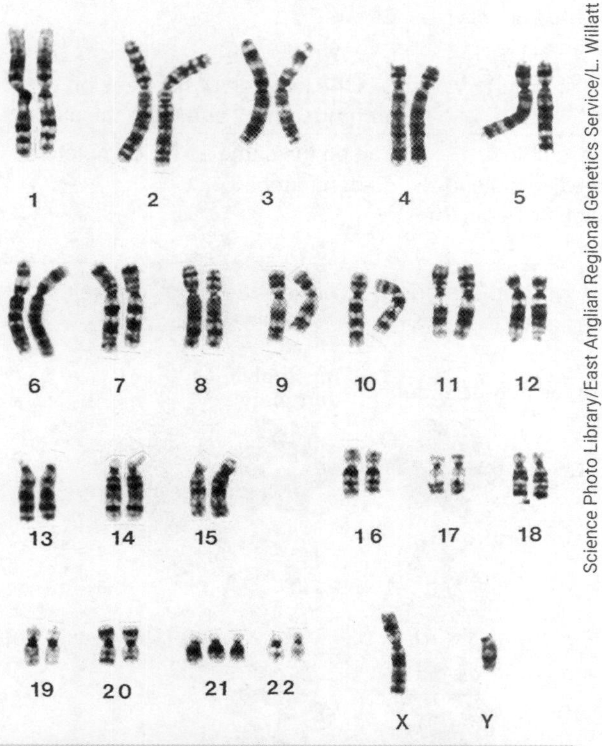

FIGURE 4.23 An additional copy of chromosome 21 (trisomy 21). This person is a male because there is one copy of the X chromosome and one copy of the Y chromosome.

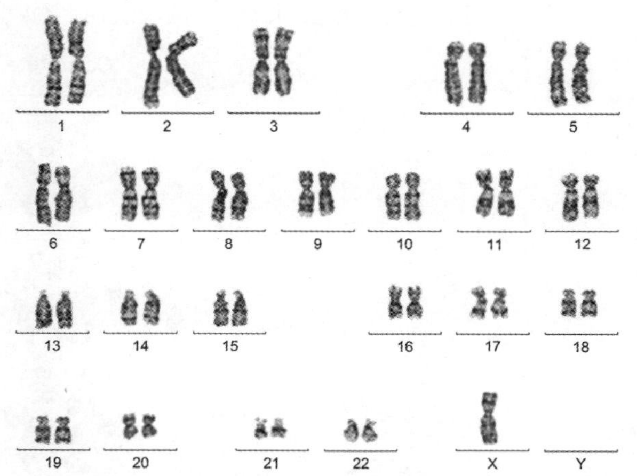

FIGURE 4.24 This karyotype has a complete set of autosomes but only one sex chromosome – X. From the information in Table 4.4, this person has Turner syndrome.

4.5 Inheritance

4.5.1 Autosomal dominant alleles

Every cell inherits two copies of each gene because of fertilisation. Only one of the two inherited genes has to be switched on and function. **Autosomal** genes occur on one of the non-sex chromosomes numbered from 1 to 22 (according to size).

The versions of the gene inherited are called alleles. If the alleles are the same, the individual is homozygous for that gene (AA or aa). If the alleles are different, the individual is heterozygous (Aa). The way the alleles interact determines which one is expressed and the resulting phenotype. Allele interactions are summarised in Table 4.5.

Hint
When choosing a letter for a trait, use a letter that has clearly different capital and lower-case forms, e.g. B or b rather than C or c.

TABLE 4.5 Allele interaction types for autosomal genes (found on non-sex chromosomes)

Trait	Description	Notation used
Complete dominance	The **dominant** allele will be expressed regardless of the other allele for this trait. It hides the presence of the other allele. AA and Aa will have the same phenotype. This is complete dominance and three allele combinations produce only two phenotypes.	Capital letter, e.g. B
	The **recessive** allele is suppressed if a dominant allele is present and will only be expressed when homozygous for this trait. aa is the only genotype that will lead to the recessive phenotype.	Lower-case letter, e.g. b
Co-dominance	For heterozygotes, each allele in the gene pair is expressed equally (*not* a blend). Three allele combinations produce three phenotypes, e.g. roan cattle colour have red *and* white hair.	C^R – red C^W – white C is the location of the gene on the chromosome. ^{R} and ^{W} are the alleles.
Incomplete dominance	For heterozygotes, the dominant allele is not completely expressed. This looks more like a blend or an intermediate phenotype e.g. a pink flower is the result of a red parent and a white parent.	Can be the same as for complete or co-dominance.

Note: some texts will use a different notation for incomplete and co-dominance. Look at the phenotypes to decide on the interaction type.

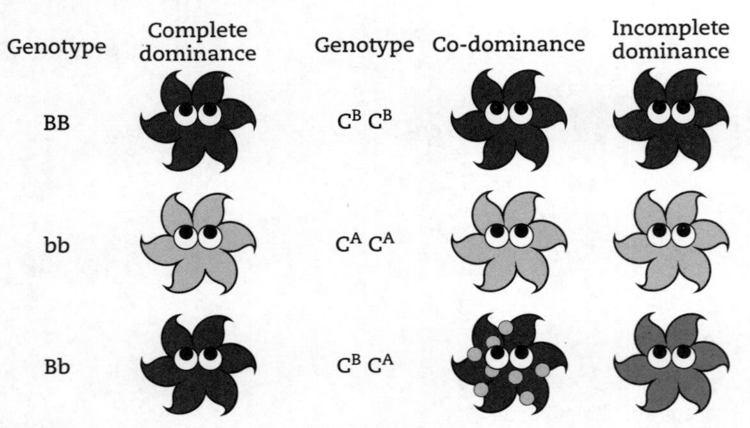

FIGURE 4.25 Types of allele interactions

Genotypes and phenotypes can be predicted with Punnett squares.

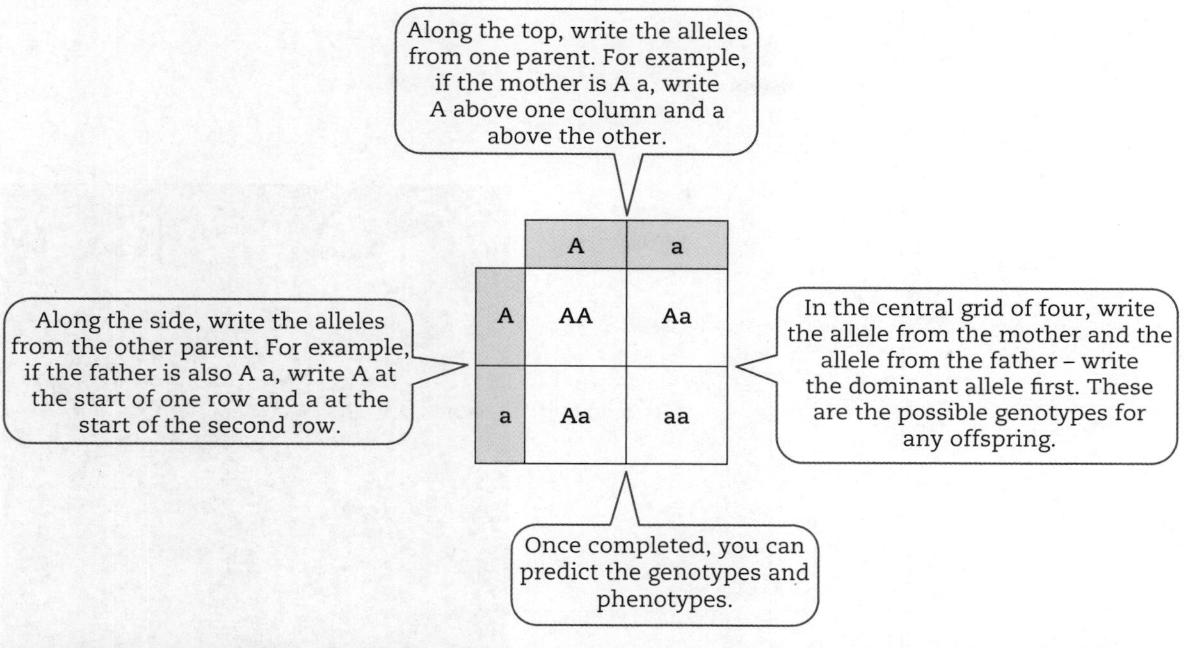

FIGURE 4.26 How to complete a Punnett square

Hint
Link this back to meiosis – the alleles in the Punnett square are determined by the possible alleles in gametes as a result of meiosis.

In the example in Figure 4.26, the genotype ratio is 1AA : 2Aa : 1aa.

Homozygous dominant: 25% or ¼

Heterozygous: 50% or ½

Homozygous recessive: 25% or ¼

The phenotype ratio is 3 dominant : 1 recessive

Dominant: 75% or ¾

Recessive: 25% or ¼

This means that each offspring from these parents has a 3 in 4 chance of having the dominant phenotype and only a 1 in 4 chance of the recessive phenotype.

4.5.2 Sex-linked traits

Sex-linked traits usually refer to genes on the X chromosome because the larger X chromosome contains more genes than the smaller Y chromosome.

What happens if the gene is on a sex chromosome?

X-linked	X-linked genes are found on the X chromosome. This means females inherit two copies and males inherit one copy (for humans). Females: the alleles can interact in any of the ways listed in Table 4.5. Males: inherit only one allele; therefore, it is expressed. No co-dominance or incomplete dominance.	X^H or X^h
Y-linked	The gene for the trait is found on the Y chromosome.	Y or –

Punnett squares are used to predict the outcomes from **sex-linked** traits too.

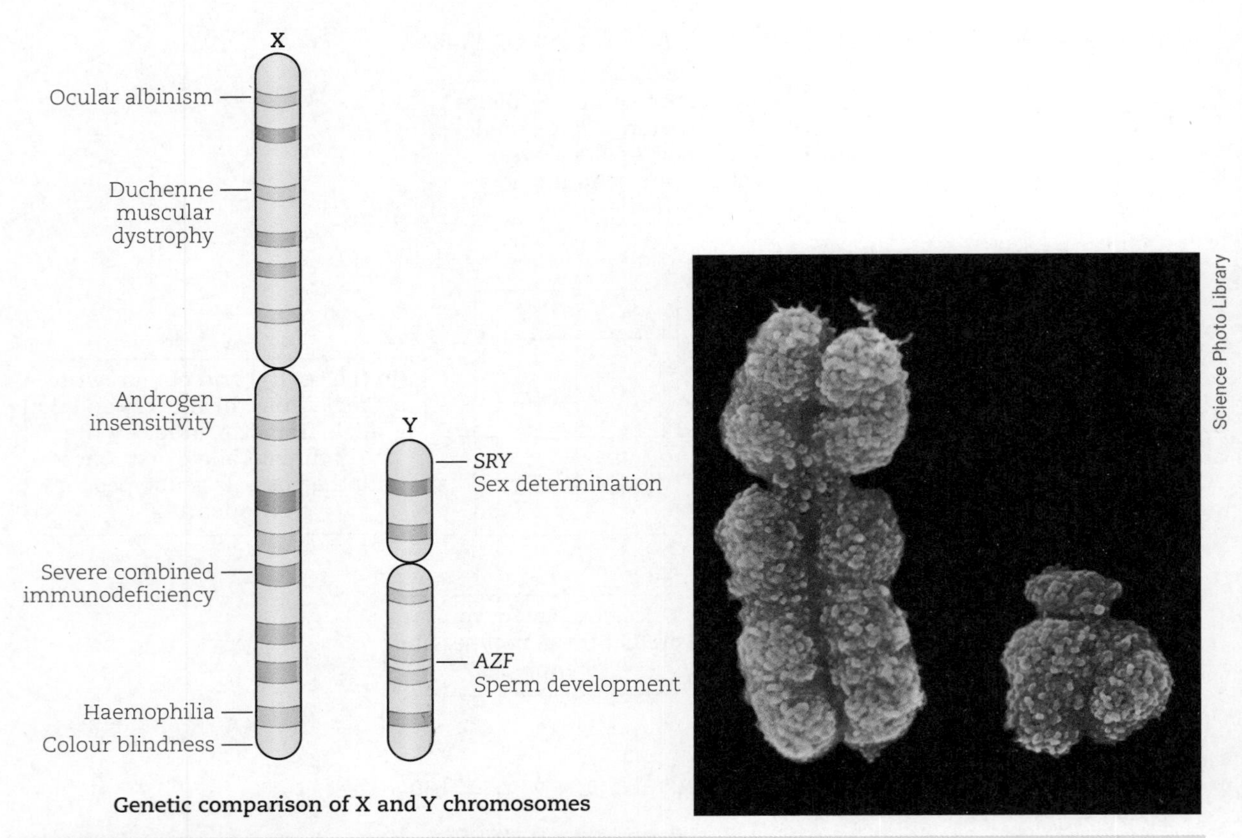

FIGURE 4.27 Genes located on the sex chromosomes. Note: the photograph shows these chromosomes as sister chromatids.

Haemophilia is a blood disorder that affects blood clotting in humans and is X-linked recessive. Because males inherit only one copy of the X chromosome, the version of the allele present will be expressed, regardless of dominance. Females inherit two copies of the X chromosome; therefore, to express the haemophilia phenotype, both X chromosomes need to have the recessive form of the allele (Figure 4.28).

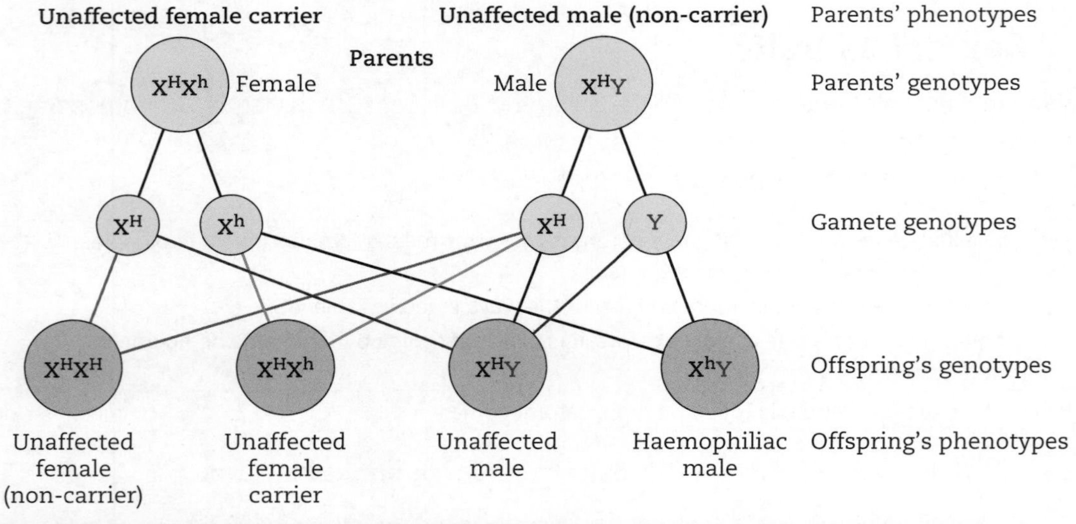

FIGURE 4.28 The inheritance pattern of haemophilia. H is the normal blood-clotting allele; h is the faulty recessive blood-clotting allele.

4.5.3 Multiple alleles

Multiple alleles occur when one gene exists in a variety of forms in a population. Although an individual may only have two copies of an allele, a population may have many different combinations of two alleles. A common example in humans is the ABO blood system, where two of the three alleles determine which proteins are expressed on the surface of red blood cells (Table 4.6 and Figure 4.29).

Hint
Note that the allele interactions described in Table 4.5 still apply!

TABLE 4.6 Possible combinations of the three alleles for blood type in humans

Blood type (phenotype)	Allele combination (genotype)
A	I^AI^A or I^Ai^o
B	I^BI^B or I^Bi^o
AB	I^AI^B
O	i^oi^o

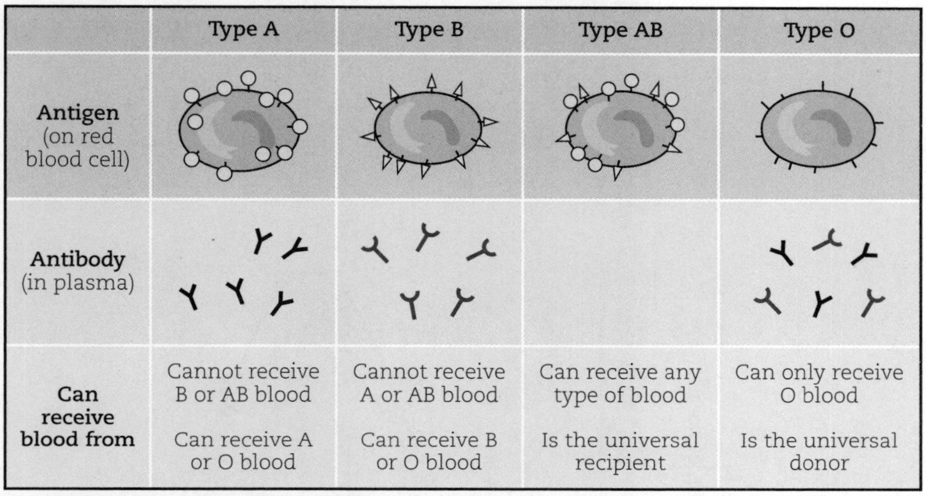

	Type A	Type B	Type AB	Type O
Antigen (on red blood cell)				
Antibody (in plasma)				
Can receive blood from	Cannot receive B or AB blood Can receive A or O blood	Cannot receive A or AB blood Can receive B or O blood	Can receive any type of blood Is the universal recipient	Can only receive O blood Is the universal donor

FIGURE 4.29 The outcome of the ABO allele interactions for blood type in humans.

4.5.4 Polygenic inheritance

Polygenic traits are more complex models of inheritance. Some features, such as height, eye colour and hair colour, are controlled by multiple genes. Two major genes and 14 other genes interact to produce all the different eye colours in humans. This can make it difficult to determine the effect of each individual gene.

Wheat grain colour is another example of a trait that follows **polygenic inheritance**. There are three genes that make a reddish pigment in wheat kernels, each of which has two alleles: A or a, B or b and D or d. The dominant allele causes a red pigment to be made. For example, the aa genotype means no pigment, the Aa genotype means some pigment, and the AA genotype means twice as much pigment as Aa.

The outcome of a cross between parents heterozygous for each of the genes is shown in Figure 4.30 (page 106), where the number of each phenotype resulting from the allele combinations is represented in the grid. This can also be represented in a frequency histogram.

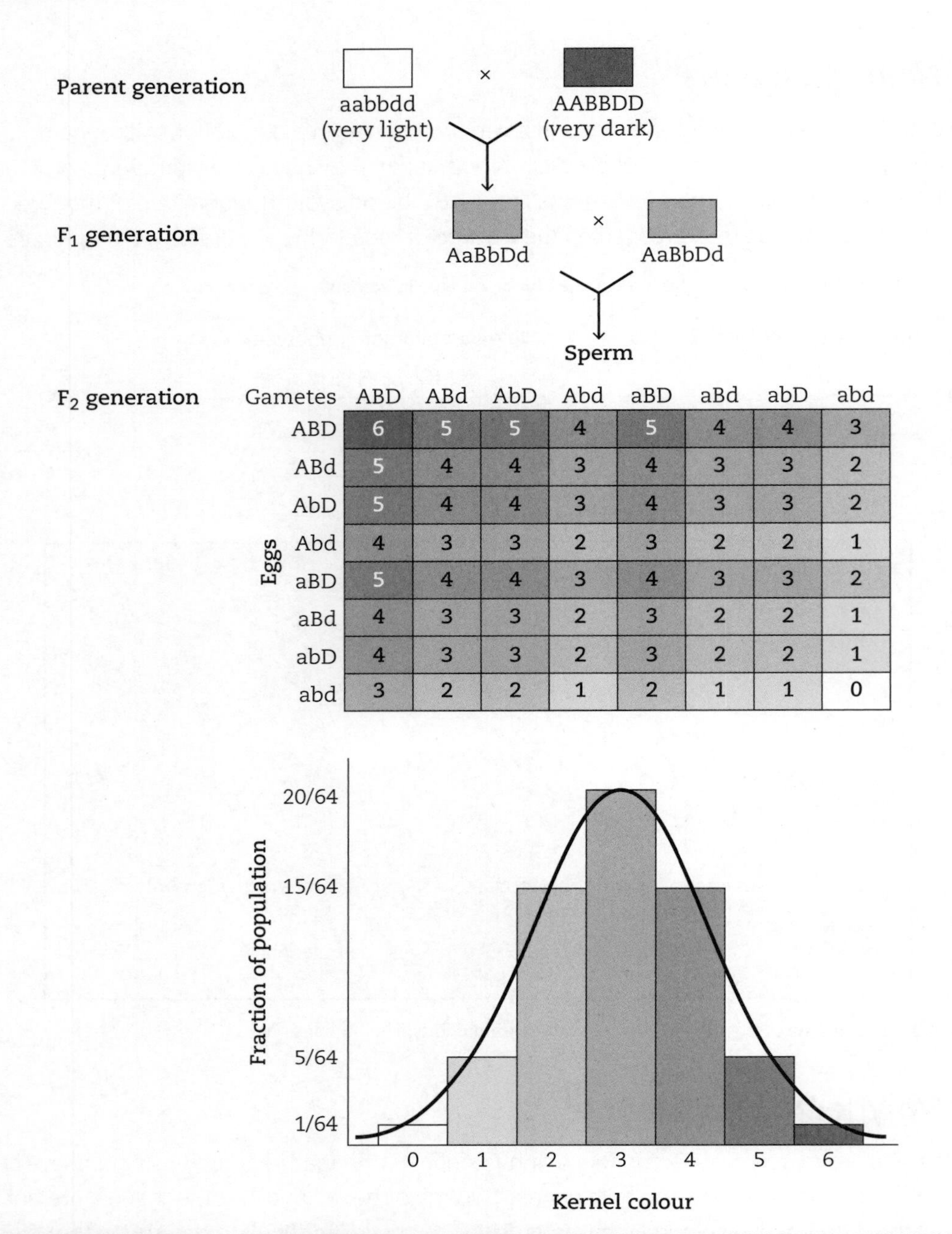

FIGURE 4.30 Possible kernel colours of wheat from a cross of heterozygous parents

The offspring have seven possible colour groups, ranging from no pigment (aabbdd) and white kernels to a lot of pigment (AABBDD) and dark red kernels. This is a form of continuous variation, as shown in the graph.

4.6 Biotechnology

4.6.1 Recombinant DNA

Recombinant DNA results when an organism's genetic material is artificially modified. It is produced when DNA from one species is inserted into the genome of a host organism of a different species to produce a transgenic organism. The resulting genetic combinations can be useful to science, medicine, agriculture and industry.

The following steps show how to produce recombinant DNA from bacteria.

1 The DNA is isolated. It is removed from a cell by breaking plasma membranes and using enzymes to destroy other molecules.

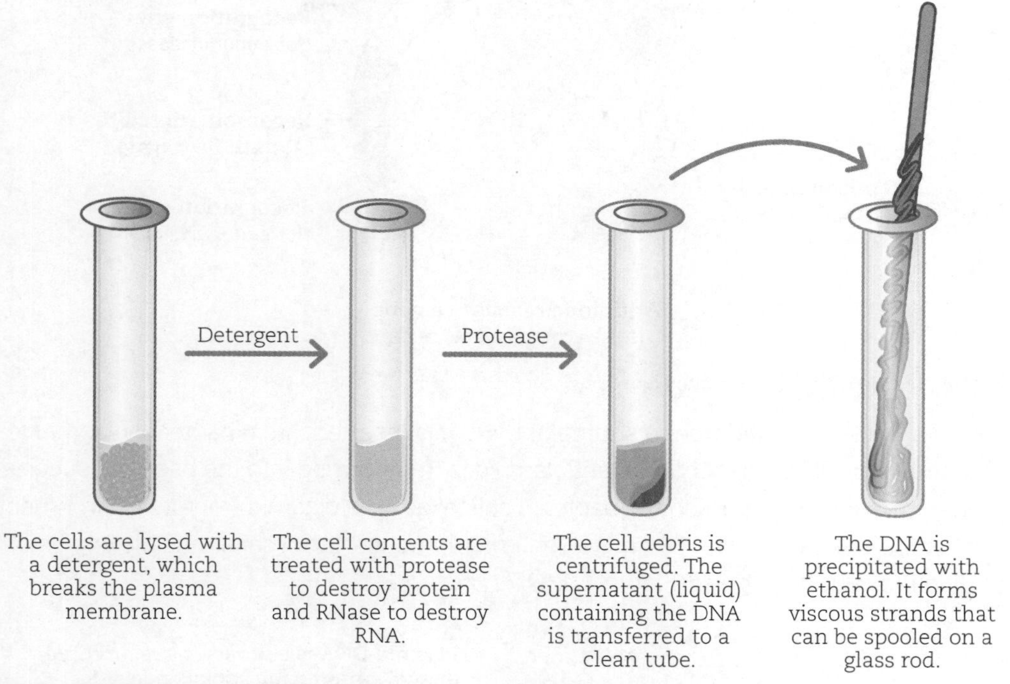

2 The DNA is cut by restriction enzymes at specific base sequences that are palindromes.

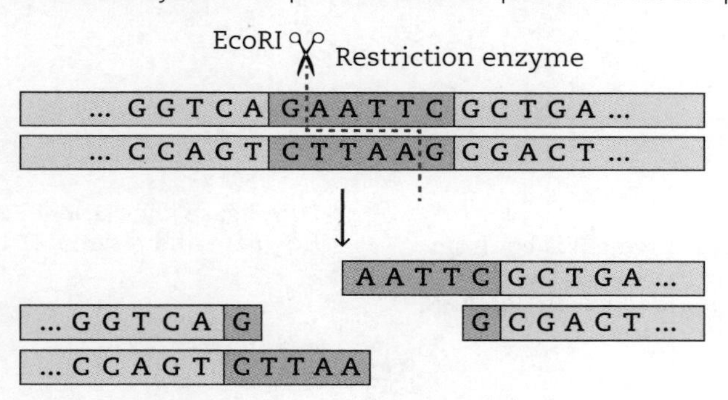

TABLE 4.7 Restriction enzyme cutting sequences

Enzyme	Target sequence	Cleavage
EcoRI	5' GAATTC 3' 3' CTTAAG 5'	5' G AATTC 3' 3' CTTAA G 5'
EcoRV	5' GATATC 3' 3' CTATAG 5'	5' GAT ATC 3' 3' CTA TAG 5'
HaeIII	5' GGCC 3' 3' CCGG 5'	5' GG CC 3' 3' CC GG 5'
HindIII	5' AAGCTT 3' 3' TTCGAA 5'	5' A AGCTT 3' 3' TTCGA A 5'
PpuMI	5' RGGWCCY 3' 3' YCCWGGR 5'	5' RG GWCCY 3' 3' YCCWG GR 5'

3 The DNA is usually inserted into a plasmid vector. A **plasmid** is a small, circular, double-stranded DNA molecule found in bacteria that is separate from the cell's chromosome. Plasmids often contain genes conferring antibiotic resistance (Figure 4.31).

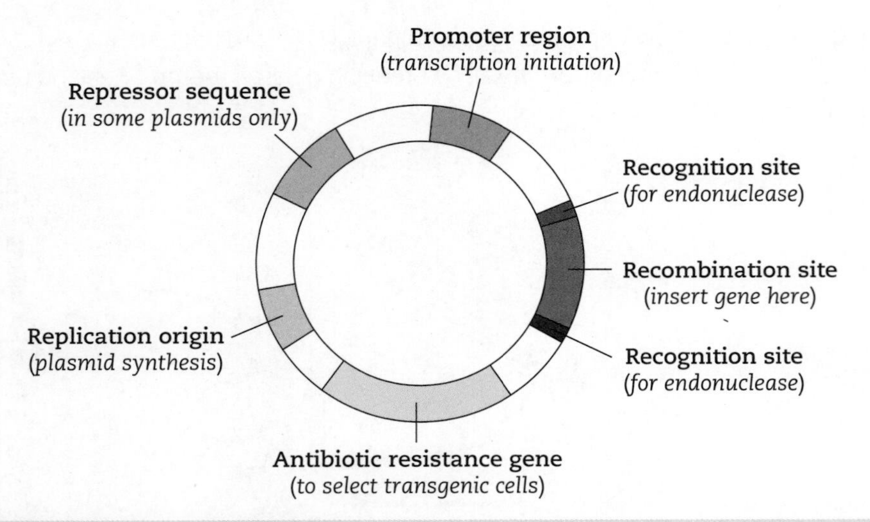

FIGURE 4.31 A plasmid with particular sites labelled

A gene of interest is isolated from a source of DNA. It is inserted into a bacterial plasmid to produce recombinant DNA. The recombinant plasmid is transformed into the bacterial species of choice. Transformation occurs when the bacterial cell takes up a plasmid from its environment directly across the cell membrane. When the bacteria reproduce, they make multiple copies of their DNA, including the gene of interest (Figure 4.32).

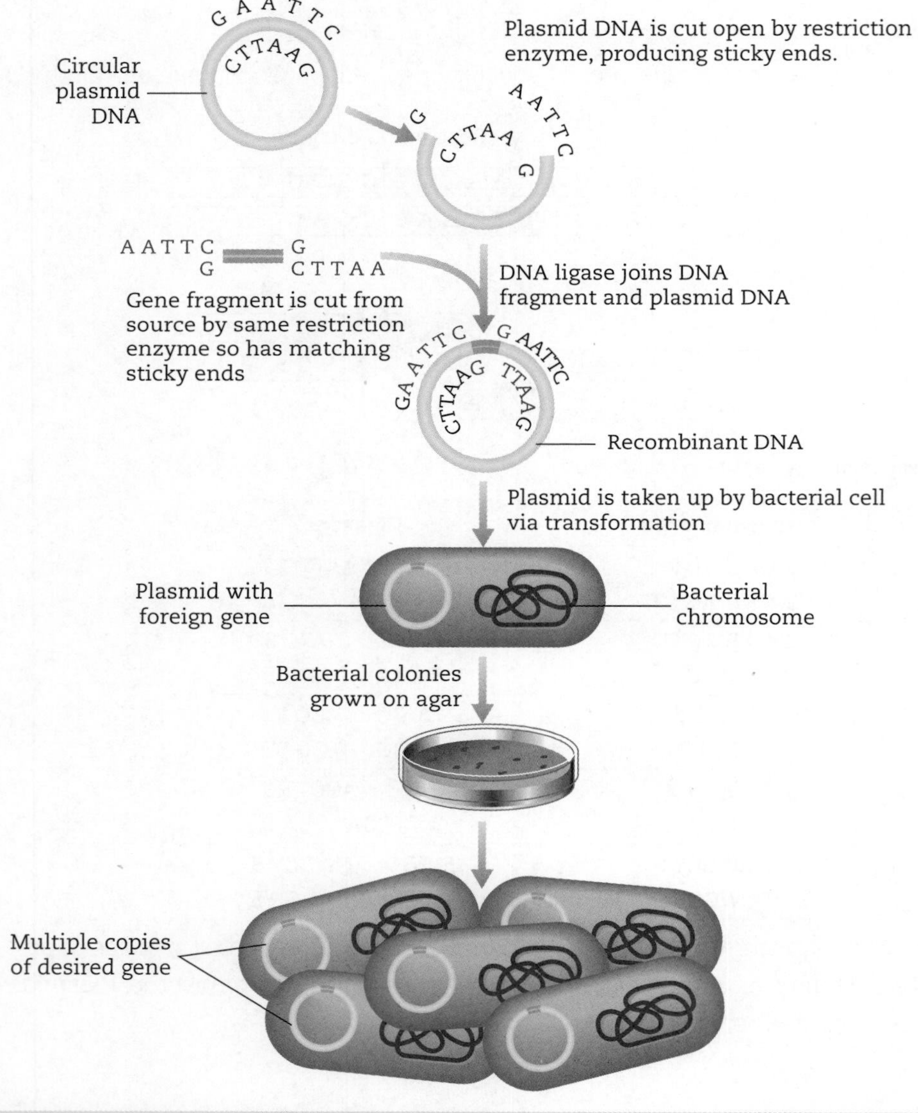

FIGURE 4.32 Steps involved in producing recombinant DNA

4.6.2 DNA sequencing

DNA sequencing is the process of determining the exact order of nucleotide bases in a molecule of DNA.

Chain termination, or Sanger sequencing, is used to make many copies of the target DNA (typically not more than 900 bp long), making fragments of different lengths. The end of each fragment has a 'chain terminator' nucleotide that fluoresces. The chain terminator marks the ends of the fragments and allows the sequence to be determined after the sequences have been organised in order of size.

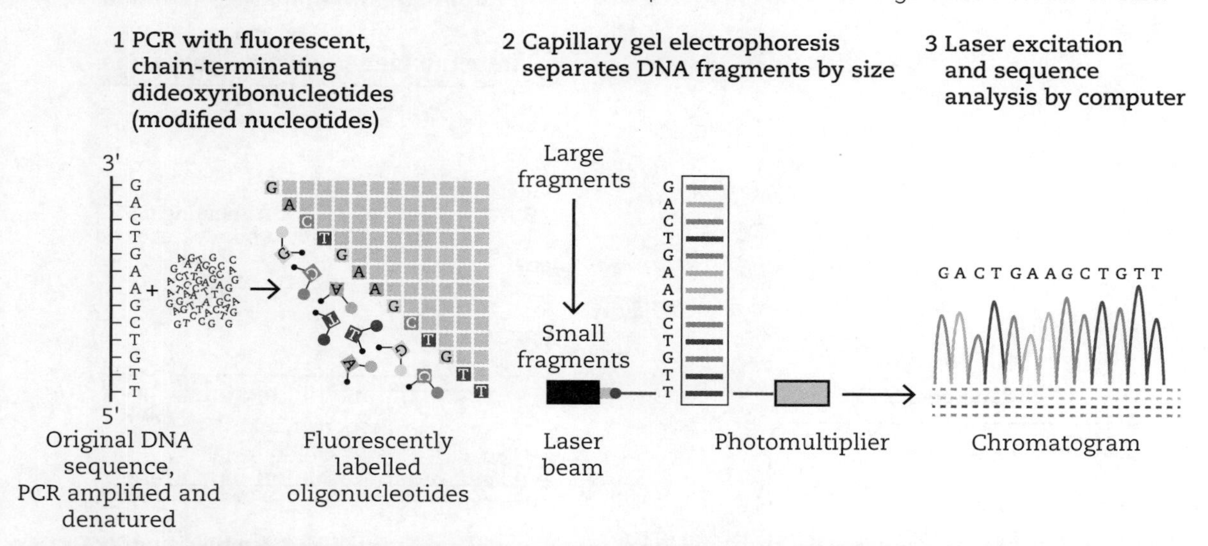

FIGURE 4.33 Sequencing of DNA by the Sanger sequencing method. Note that the method also uses aspects of the polymerase chain reaction and gel electrophoresis.

There are also 'next generation sequencing' processes. Next generation sequencing (NGS) refers to a collection of DNA and RNA sequencing technologies that allow researchers to sequence entire genomes or specific sequences within genomes much faster and at a lower cost. Also known as 'massively parallel sequencing', these techniques allow for simultaneous sequencing and analysis of multiple genes or nucleotide samples.

4.6.3 Polymerase chain reaction

The polymerase chain reaction (PCR) is a process that uses a thermocycler to amplify or copy sections of DNA to ensure there is sufficient DNA for molecular and genetic analyses; for example, DNA fingerprinting, viral identification or identification of genetic disorders. There are three main steps:

1. Denaturing the DNA at a high temperature (95°C) to separate the strands
2. Annealing – adding short DNA primers at 55°C
3. Extension or synthesising – using Taq DNA polymerase (from the bacteria *Thermus aquaticus*) and nucleotides to build complementary DNA strands at 72°C.

These steps are repeated many times to produce many copies of the required DNA sequence (Figure 4.34, page 108).

9780170459136

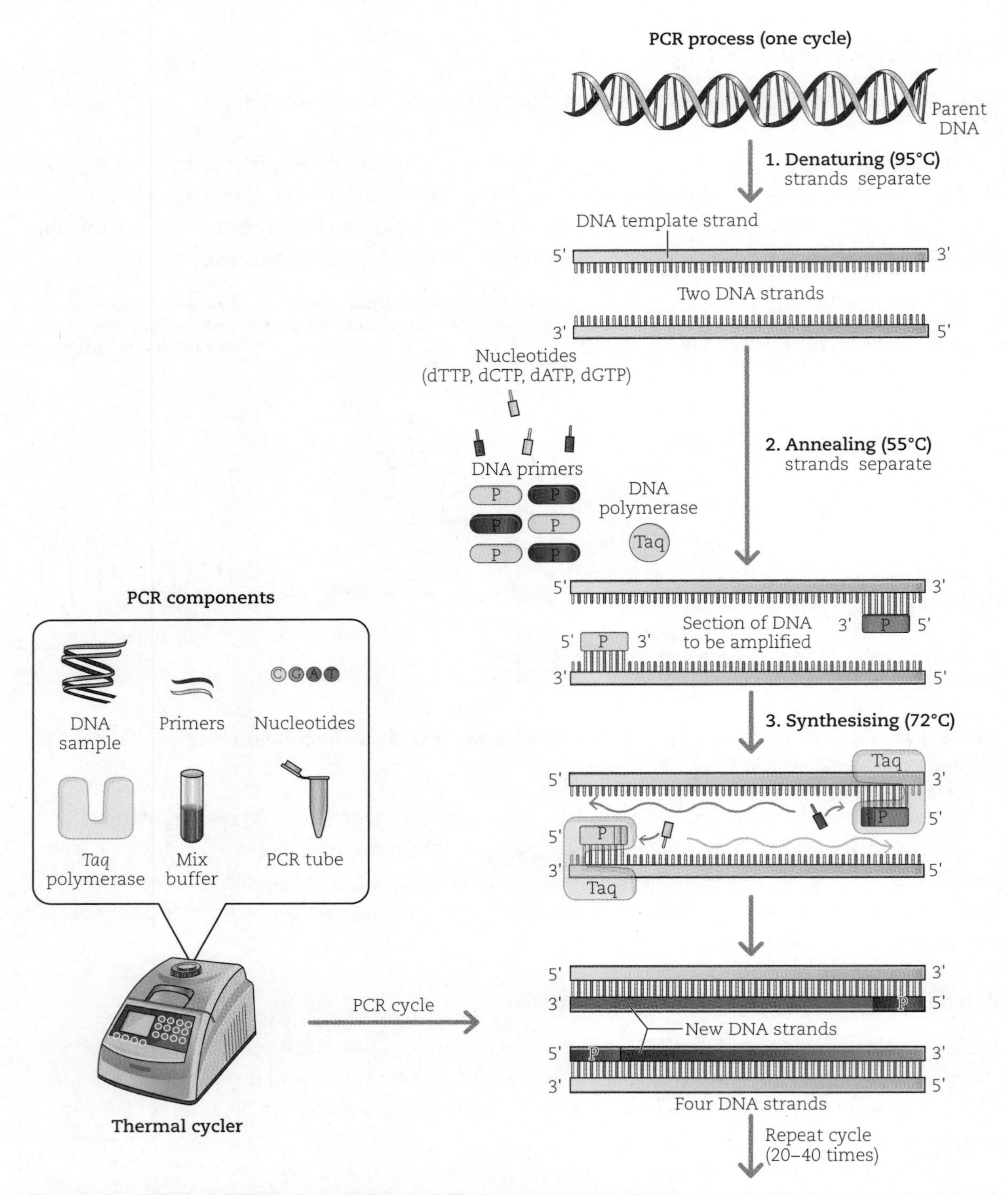

FIGURE 4.34 The three steps involved in PCR: denaturing, annealing and synthesising

4.6.4 Gel electrophoresis

Gel electrophoresis is used to separate molecules of DNA, RNA or proteins on the basis of size (length and shape) and charge. Fragments with a charge can be pulled through a gel by an electric field. The smaller the molecule, the faster it moves through the gel towards the oppositely charged electrode. DNA has a negative charge and therefore moves towards the positive electrode.

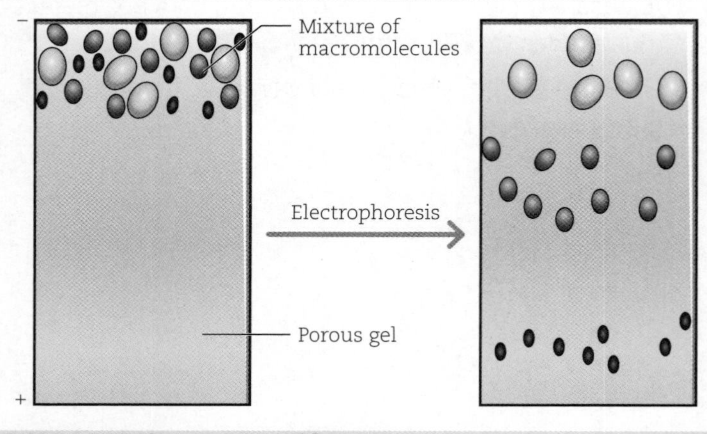

FIGURE 4.35 The separation of fragments

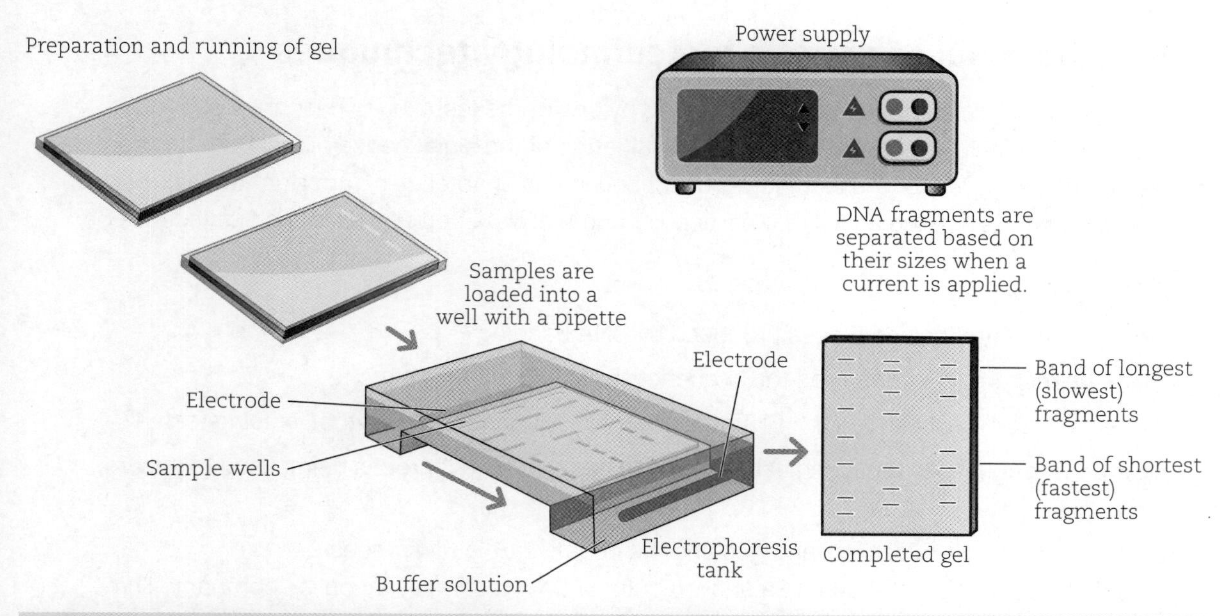

FIGURE 4.36 The stages of gel electrophoresis. After DNA has been extracted, DNA fragment samples are usually treated with restriction enzymes to prepare the sample.

The result is a series of bands of fragments of the same length at one point in the gel or a DNA band. The gel usually has a run of fragments of known length to produce a 'ladder' (or molecular marker), that the other bands are compared with. The bands can be seen because they are labelled with fluorescent tags.

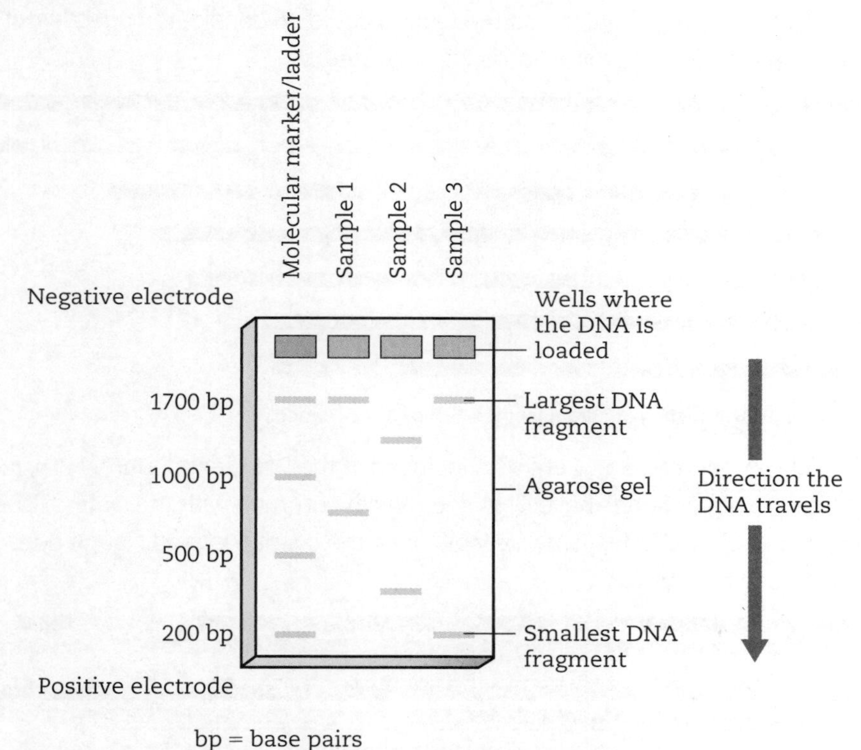

FIGURE 4.37 The results of a gel electrophoresis. Molecular markers of known size are run alongside samples and can be used to estimate the size of the DNA fragments migrating through the gel (bp = base pairs).

Uses for electrophoresis include:

- forensics
- paternity testing
- evolutionary studies.

4.6.5 Appraisal of genetic biotechnology techniques

The syllabus requires appraisal of data produced by one of the techniques described above. As described in Chapter 2, appraisal requires the evaluation of the significance or worth, in this case of data produced by any of the techniques described in this chapter. It is important to decide on what the 'success rate' is defined as for that particular source of data before evaluating significance or worth.

Options to consider:

- How accurate are the sequences produced by DNA banding?
- Have the researchers confirmed the sequences?
- Is there a clear 'link' between the source of forensic DNA and the DNA that is being tested?
- How many sequences are being assessed? Are there sufficient probes being used to ensure a match?

Look at both the positives/advantages and the negatives/disadvantages.

For example, Figures 4.38 and 4.39 show results of DNA testing by an online company. Three women (dark grey, mid grey, light grey on the grid) have their DNA tested to get more information about their ancestry. Multiple DNA markers are analysed and the results for chromosomes 1–7 are presented on Figure 4.38. The dark grey (second woman) over the light colour (first woman) shows where there is significant overlap between two of the women, indicating they have significant commonality in their DNA. The third woman has no overlap. It could be concluded that she is unrelated.

In fact, the first two women are sisters, so the testing could be considered successful because it has identified their relationship. The third woman is unrelated.

FIGURE 4.38 The results of DNA testing of chromosomes 1–7 of three women

However, when chromosomes 8–14 are also included in the DNA test (Figure 4.39), it can be seen that the third woman (mid grey) does have DNA sections in common with the sisters. Does that make the third woman a sister as well? This is unlikely because the overlap is small, but it does suggest that she has some ancestors in common.

FIGURE 4.39 The results of DNA testing of chromosomes 8–14 of the same three women as in Figure 4.38

9780170459136

Analysis of this data suggests that two women are more closely related to each other than they are to the third woman. But it was only possible to determine this once enough data was analysed (i.e. chromosomes 1–14). Chromosomes 1–7 did not provide enough information to draw this conclusion.

Glossary

A+ DIGITAL FLASHCARDS
Revise this topic's key terms and concepts by scanning the QR code or typing the URL into your browser.

https://get.ga/aplus-qce-bio-u34

Allele
An alternative form of a gene that arises through mutation.

Aneuploidy
The result of an extra or missing chromosome in a gamete.

Anticodon
The complement of a codon, e.g. AAA is the complement of TTT.

Autosome
A non-sex chromosome found in the nucleus of a cell.

Catalyse
Make a reaction proceed or speed up by adding a catalyst (e.g. proteins).

Characteristic
A distinguishing feature in an organism.

Chromosome
A long structure in the nucleus made of DNA that carries genetic information.

Chromatid
One half of a duplicated chromosome produced during DNA replication.

Codon
The sequence of three nucleotides that is the basis of the genetic code in DNA and RNA.

Complementary
In DNA, nucleotides pair up, e.g. adenosine with thymine, enabling DNA to replicate.

Diploid
Containing two copies of each chromosomes (e.g. body cells), represented as $2n$.

Dominant
The trait expressed even if only one copy of its gene is present.

Exon
A section of the chromosome that codes for proteins.

Fertilisation
The fusion of a male gamete and a female gamete to produce a zygote.

Gamete
The haploid product of meiosis: can be male or female; fusion results in a zygote.

Gene
A region(s) of DNA that is made up of nucleotides; the molecular unit of heredity.

Genome
All the genetic material in the chromosomes of an organism, including its genes and DNA sequences.

Haploid
Having half the normal number of chromosomes; represented as n.

Histone
A type of protein that binds to DNA to allow the DNA to coil to form nucleosomes.

Homologous
Chromosomes carrying the same genes that pair up during meiosis.

Intron
A section of a chromosome that does not code for proteins but is essential for gene regulation.

Karyotype
The number and appearance of chromosomes in the nucleus of an organism.

Maternal
Originating from the female line.

Monomer
A simple molecule that can form covalent bonds with like molecules to form a large molecule of repeating units called a polymer.

Mutation
A change in DNA sequence that alters the sequence of amino acids produced.

Non-disjunction
The failure of homologous chromosomes to separate during meiosis; leads to aneuploidy.

Nucleotide
The basic unit that is repeated to make nucleic acids; contains a sugar, phosphate group and nitrogenous base.

Paternal
Originating from the male line.

Plasmid
A small, circular, double-stranded DNA molecule found in bacteria that is separate from the cell's chromosome.

Polygenic inheritance
When one characteristic is controlled by two or more genes.

Polypeptide
A polymer made up of amino acids.

Recessive
The trait expressed only if two copies of the gene are present.

Recombinant DNA
A DNA molecule formed when DNA from different species are combined.

Sex-linked
Traits determined by genes on the X chromosome.

Trait
See *characteristic*.

Variation
Differences between cells and organisms caused by genetics and/or the environment.

Zygote
The result of the fusion of a male sex cell and a female sex cell, i.e. a fertilised egg.

Revision summary

Use the following summary of syllabus dot points and key knowledge within Unit 4 Topic 1 to ensure that you have thoroughly reviewed the content. Provide a brief definition or comment for each item to demonstrate your understanding or code them using the traffic light system – green (all good), amber (needs some review), red (priority area to review).

DNA structure and replication	
• understand that deoxyribonucleic acid (DNA) is a double-stranded molecule that occurs bound to proteins (histones) in chromosomes in the nucleus, and as unbound circular DNA in the cytosol of prokaryotes, and in the mitochondria and chloroplasts of eukaryotic cells	
• recall the structure of DNA, including – nucleotide composition – complementary base pairing – weak, base-specific hydrogen bonds between DNA strands	
• explain the role of helicase (in terms of unwinding the double helix and separation of the strands) and DNA polymerase (in terms of formation of the new complementary strands) in the process of DNA replication. Reference should be made to the direction of replication.	
Cellular replication and variation	
• within the process of meiosis I and II – recognise the role of homologous chromosomes – describe the processes of crossing over and recombination and demonstrate how they contribute to genetic variation – compare and contrast the process of spermatogenesis and oogenesis (with reference to haploid and diploid cells).	
• demonstrate how the process of independent assortment and random fertilisation alter the variations in the genotype of offspring.	
Gene expression	
• define the terms *genome* and *gene*	

››

››

• understand that genes include 'coding' (exons) and 'noncoding' DNA (which includes a variety of transcribed proteins: functional RNA (i.e. tRNA), centromeres, telomeres and introns. Recognise that many functions of 'noncoding' DNA are yet to be determined)	
• explain the process of protein synthesis in terms of – transcription of a gene into messenger RNA in the nucleus – translation of mRNA into an amino acid sequence at the ribosome (refer to transfer RNA, codons and anticodons)	
• recognise that the purpose of gene expression is to synthesise a functional gene product (protein or functional RNA); that the process can be regulated and is used by all known life	
• identify that there are factors that regulate the phenotypic expression of genes – during transcription and translation (proteins that bind to specific DNA sequences) – through the products of other genes – via environmental exposure (consider the twin methodology in epigenetic studies)	
• recognise that differential gene expression, controlled by transcription factors, regulates cell differentiation for tissue formation and morphology	
• recall an example of a transcription factor gene that regulates morphology (HOX transcription factor family) and cell differentiation (sex-determining region Y).	
Mutations	
• identify how mutations in genes and chromosomes can result from errors in – DNA replication (point and frameshift mutation) – cell division (non-disjunction) – damage by mutagens (physical, including UV radiation, ionising radiation and heat and chemical)	
• explain how non-disjunction leads to aneuploidy	

››

• use a human karyotype to identify ploidy changes and predict a genetic disorder from given data	
• describe how inherited mutations can alter the variations in the genotype of offspring.	
Inheritance	
• predict frequencies of genotypes and phenotypes using data from probability models (including frequency histograms and Punnett squares) and by taking into consideration patterns of inheritance for the following types of alleles: autosomal dominant, sex linked and multiple	
• define *polygenic inheritance* and predict frequencies of genotypes and phenotypes for using three of the possible alleles.	
Biotechnology	
• describe the process of making recombinant DNA – isolation of DNA, cutting of DNA (restriction enzymes) – insertion of DNA fragment (plasmid vector) – joining of DNA (DNA ligase) – amplification of recombinant DNA (bacterial transformation)	
• recognise the applications of DNA sequencing to map species' genomes and DNA profiling to identify unique genetic information	
• explain the purpose of polymerase chain reaction (PCR) and gel electrophoresis	
• appraise data from an outcome of a current genetic biotechnology technique to determine its success rate.	

Exam practice

Multiple-choice questions

Each multiple-choice question is worth 1 mark.

Solutions start on page 162.

The following diagram relates to Questions 1 and 2.

A simplified result of crime scene DNA analysis of blood

Question 1

The shortest DNA fragment in the crime scene lane is fragment

A A

B B

C C

D D

Question 2

On the basis of these results, it is accurate to say that the crime scene sample

A does not contain blood from one of the suspects.

B contains blood from each of the suspects.

C contains blood from suspect 2.

D contains blood from suspect 3.

Question 3

A nucleotide consists of

A sugar, base, phosphate.

B sucrose, base, phosphate.

C amino acid, base, phosphate.

D amino acid, base, phosphorus.

Question 4

Complementary bases in DNA are held together by

A covalent bonds.

B hydrogen bonds.

C ionic bonds.

D metallic bonds.

Question 5

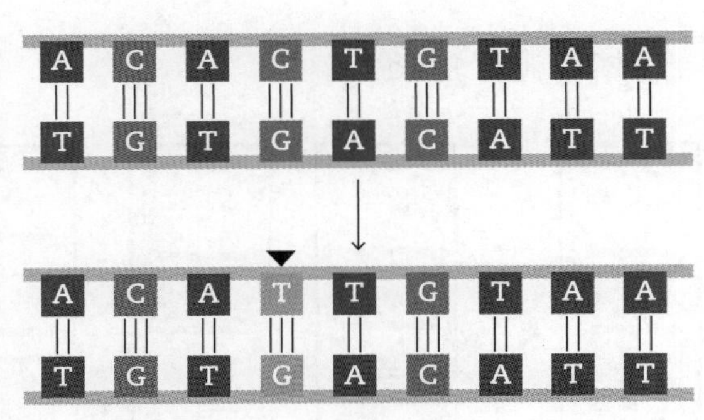

The diagram shows an example of

A deletion.

B insertion.

C non-disjunction.

D substitution.

Question 6

A white blood cell and a nerve cell from one person contain the same genetic sequences. However, the nerve cell and white blood cell carry out very different roles. This is possible because each cell has

A different genes.

B different genes activated.

C different DNA sequences.

D different types of mutations.

Question 7 ©QCAA QCAA 2020 P1 MC Q10

DNA profiling using polymerase chain reaction (PCR) and gel electrophoresis allows the comparison of

A genes.

B entire genomes.

C DNA fragments.

D specific sites of mutations.

Question 8 ©QCAA QCAA 2020 P1 MC Q15

In watermelon, skin colour is controlled by a single autosomal gene. The two phenotypic variants are green and striped. Two plants, one homozygous for the green alleles, and one homozygous for the striped alleles, were crossed. The figure shows the phenotypic frequency for the initial (F_0) generation and the subsequent (F_1) generation.

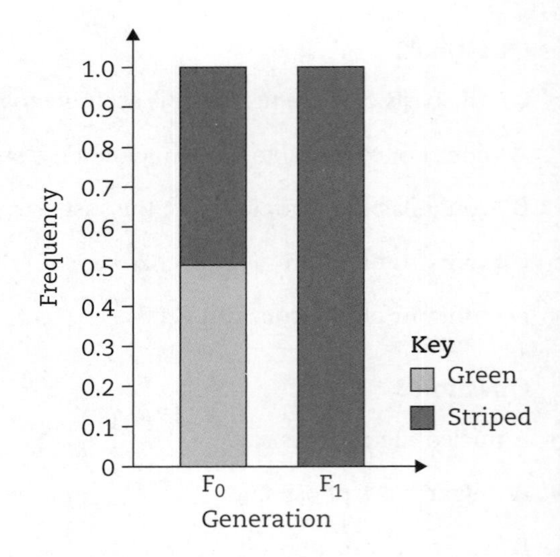

A cross was then performed between members of the F_1 generation. What would be the expected frequency of striped watermelon in the next (F_2) generation?

A 1.0

B 0.75

C 0.50

D 0.25

Short-response questions

Solutions start on page 163.

Question 9 (6 marks)

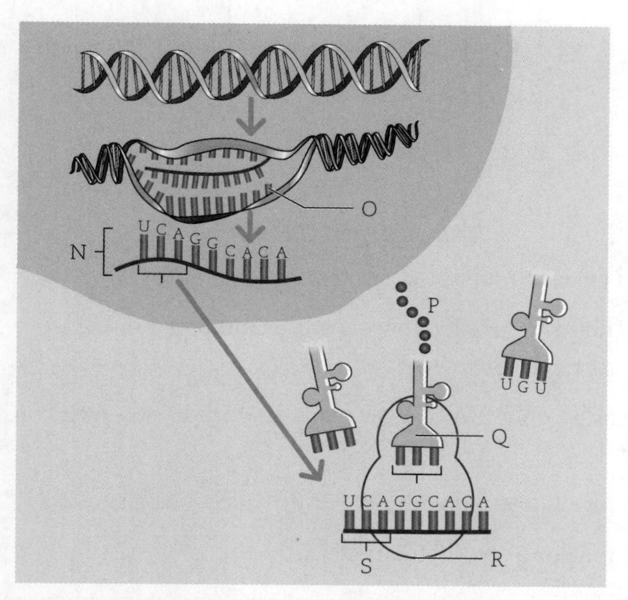

Explain the process of protein synthesis by completing the annotations for this diagram for N–S.

Question 10 (14 marks)

Researchers are investigating plasmids for use in making recombinant DNA. The restriction sites for EcoRI and HindIII are marked.

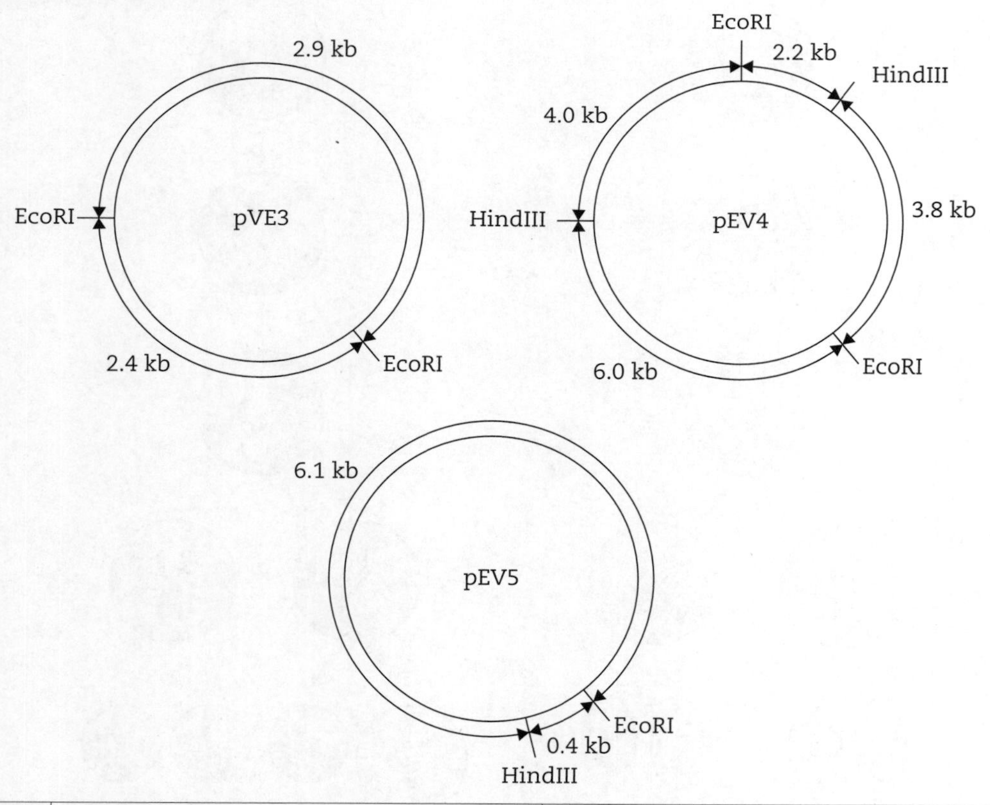

HindIII	A↓AGCTT TTCGA↑A	EcoRI	G↓AATTC CTTAA↑G

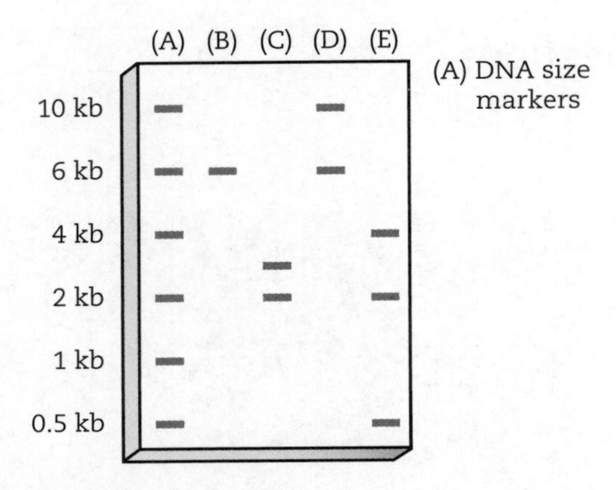

a Explain the purpose of gel electrophoresis. 2 marks

b Plasmid pVE6 (not pictured) is lane E. How big is plasmid pVE6 in kb? Use the data in the agarose gel. 1 mark

c pVE6 was digested by using only EcoRI. How many restriction sites for EcoRI are there on this plasmid? 1 mark

d Which channel shows pEV4 digested with EcoRI? Provide a reason. 3 marks

e Describe the process of making recombinant DNA. 4 marks

f The target gene for producing a recombinant plasmid includes this sequence: AAGCTT. Which plasmid is the best plasmid to use for this transformation to be successful? Provide reasons for your answer. 3 marks

Question 11 (5 marks)

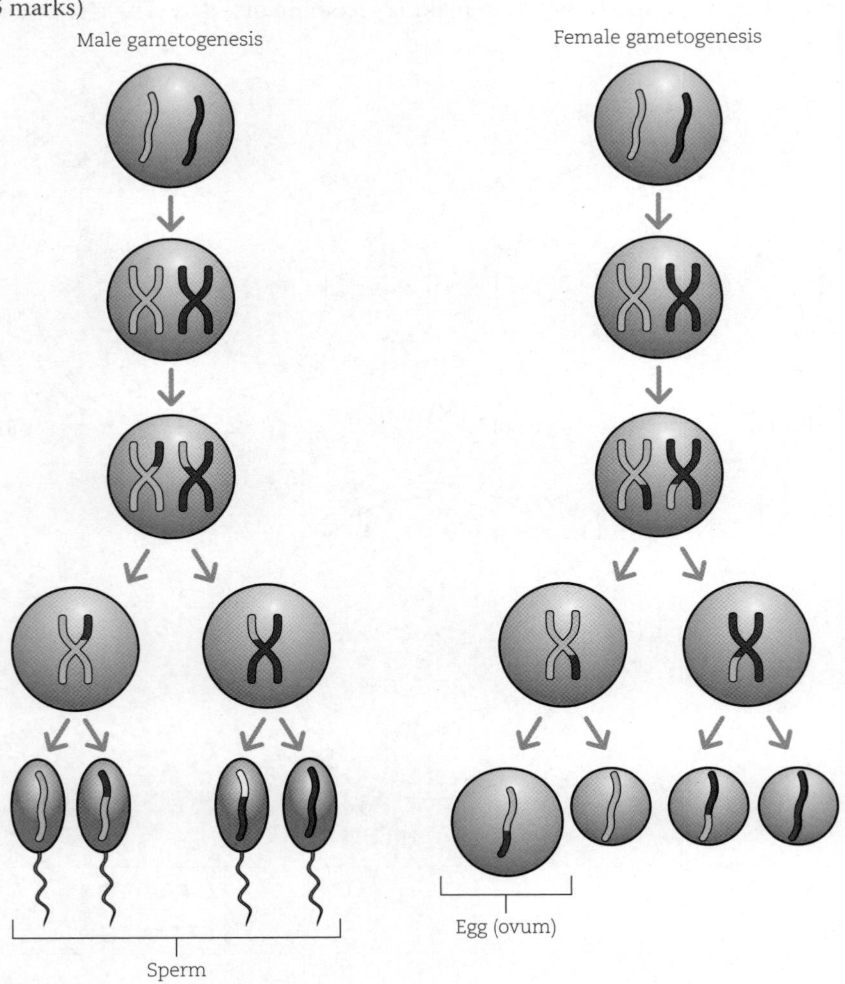

a On the diagram, circle an example of homologous chromososmes. 1 mark

b Identify one similarity and one difference between spermatogenesis and oogenesis in the diagram. Explain the significance of each. 4 marks

Question 12 (9 marks)

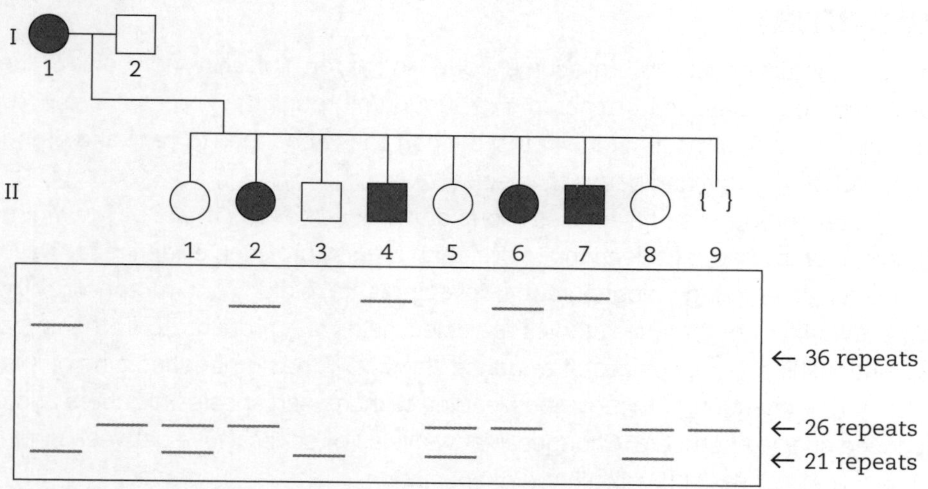

A genetic disorder results from multiple repeats (>36) of a CAG sequence in the DNA. Data from a gel electrophoresis is shown above, along with a pattern of inheritance for a family. Circles represent females and squares males. Coloured-in shapes represent individuals who have inherited the disorder.

a Is this disorder autosomal or X-linked? Provide a reason for your answer. 2 marks

b How many individuals in generation II are affected by this disorder? 1 mark

c Predict whether individual 9 will have the disorder. Provide a reason. 2 marks

d Use a Punnett square to explain how the individuals in generation II acquired this disorder. 4 marks

Question 13 (4 marks) ©QCAA QCAA 2020 P2 SR Q3

Explain the role of the enzymes helicase and DNA polymerase in the process of DNA replication.

Question 14 (4 marks) ©QCAA QCAA 2020 P2 SR Q5

Explain the process of protein synthesis in terms of transcription and translation.

Chapter 5
Topic 2 Continuity of life on Earth

Topic summary

Genes in our cells contain genetic information encoded in the form of DNA molecules in the nucleus. The genetic information is passed from parents to offspring through the process of reproduction. Spontaneous changes in DNA occur at low rates, but can cause changes to cells and organisms and are a source of variation within and between species.

Variation between individuals of a species is one of the observations made by Darwin that led to the theory of evolution by natural selection. Darwin (and others) provided evidence for 'descent with modification' from palaeontology, biogeography, developmental biology and morphology. Advances in technology allowed for comparative studies of nucleic acids and proteins, to further support evolutionary relationships. These pieces of evidence show how the number and type of species present on Earth have changed, with radiations leading to increased species numbers and extinctions leading to reduced species numbers. Changes in how alleles of genes move between populations can be observed and used to predict the likelihood of speciation.

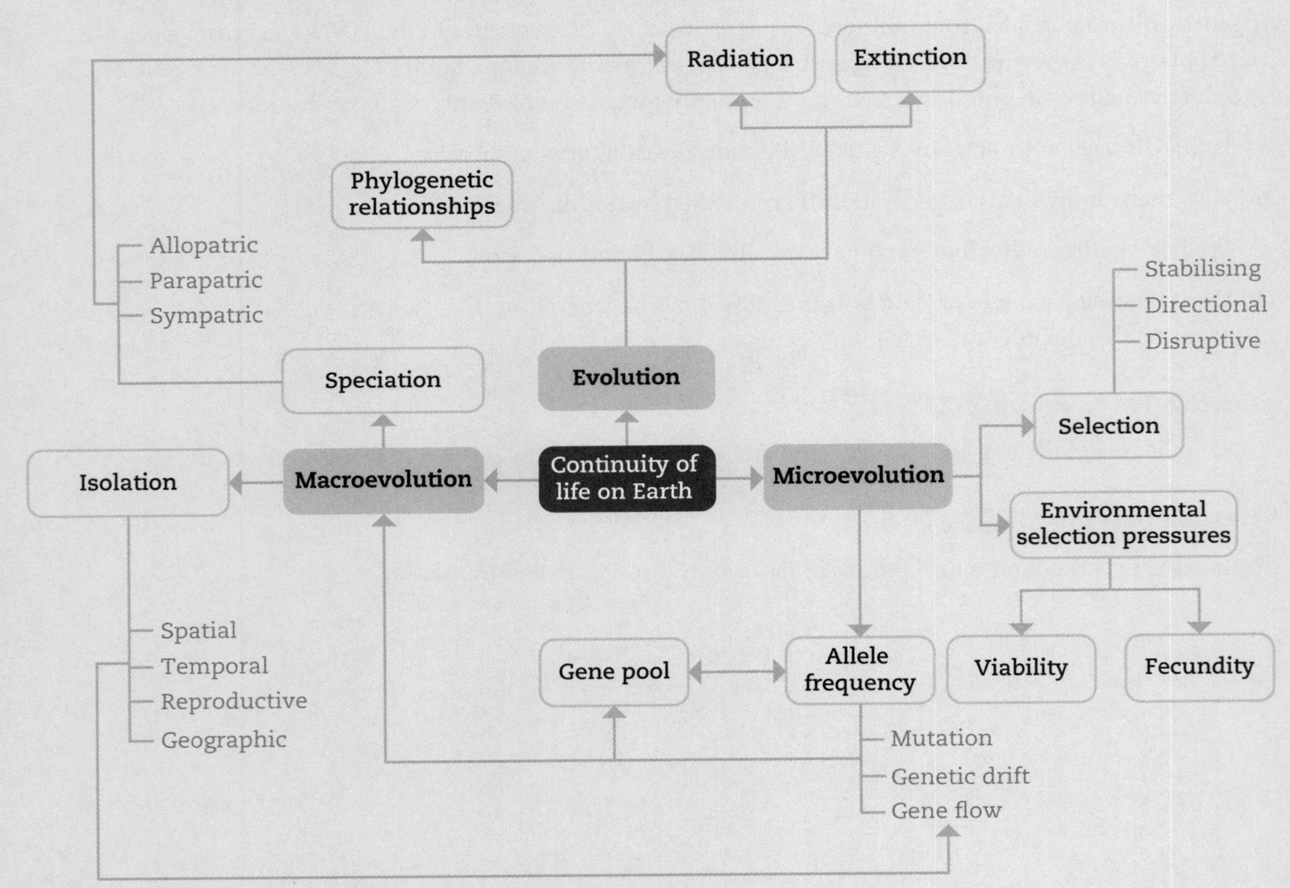

5.1 Evolution

5.1.1 Scale of evolution

Evolution is a change in the genetic composition of a population over successive generations, which may result in the development of new species. **Macroevolution** is the variation of allele frequencies at or above the level of species over geological time, resulting in the divergence of taxonomic groups, in which the descendant is in a different taxonomic group from the ancestor. **Microevolution** is small-scale variation of allele frequencies within a species or population, in which the descendant is in the same taxonomic group as the ancestor.

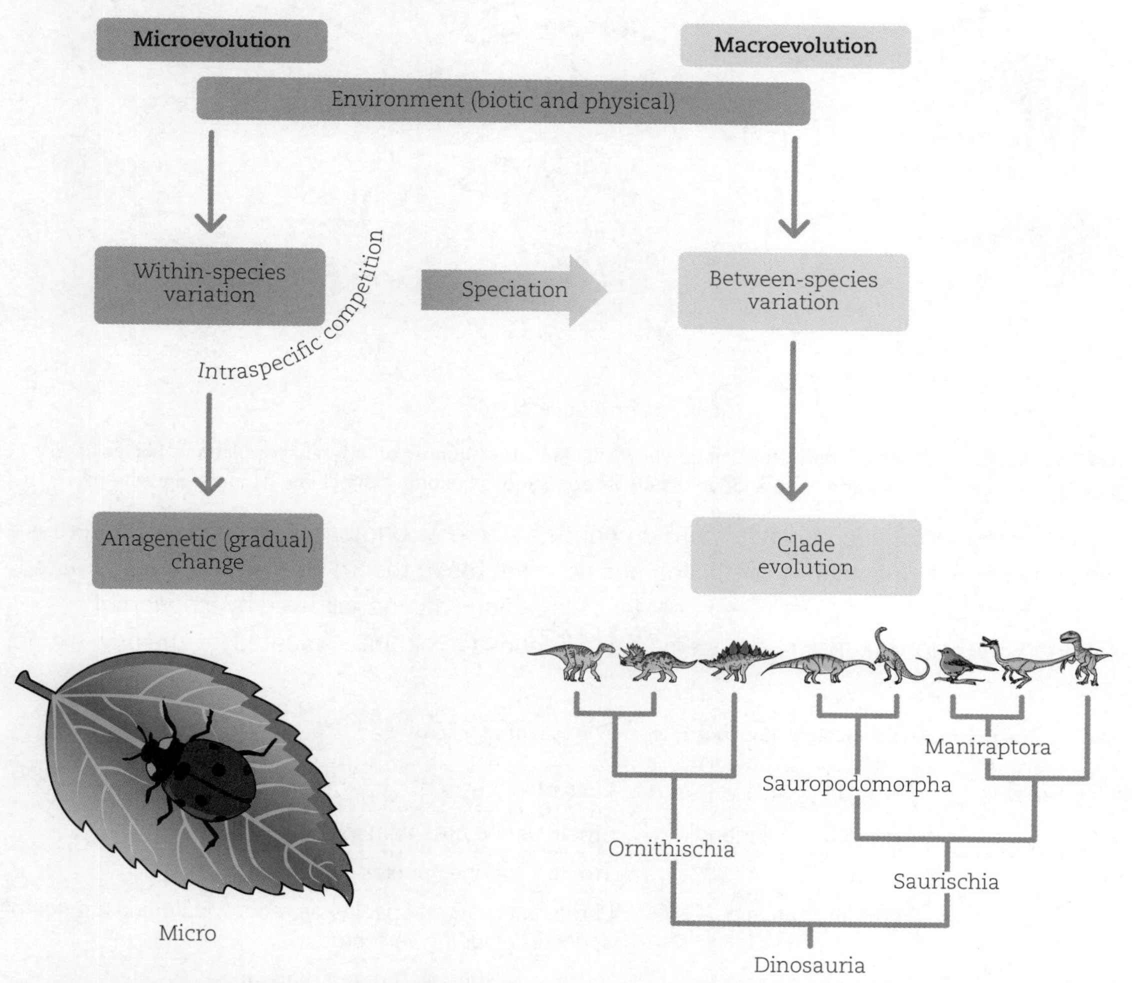

FIGURE 5.1 Examples of microevolution and macroevolution

An evolutionary **radiation** occurs when there is an increase in taxonomic diversity – there are more species in a clade than there used to be – due to a high rate of speciation. Adaptive radiation is the most frequently described type of radiation and is triggered by biotic factors. This results in a **divergence** in species **morphology** (appearance) and habitat preferences. Therefore, these species compete with each other less over time. Geographic radiation results from geographic **isolation** leading to speciation.

9780170459136

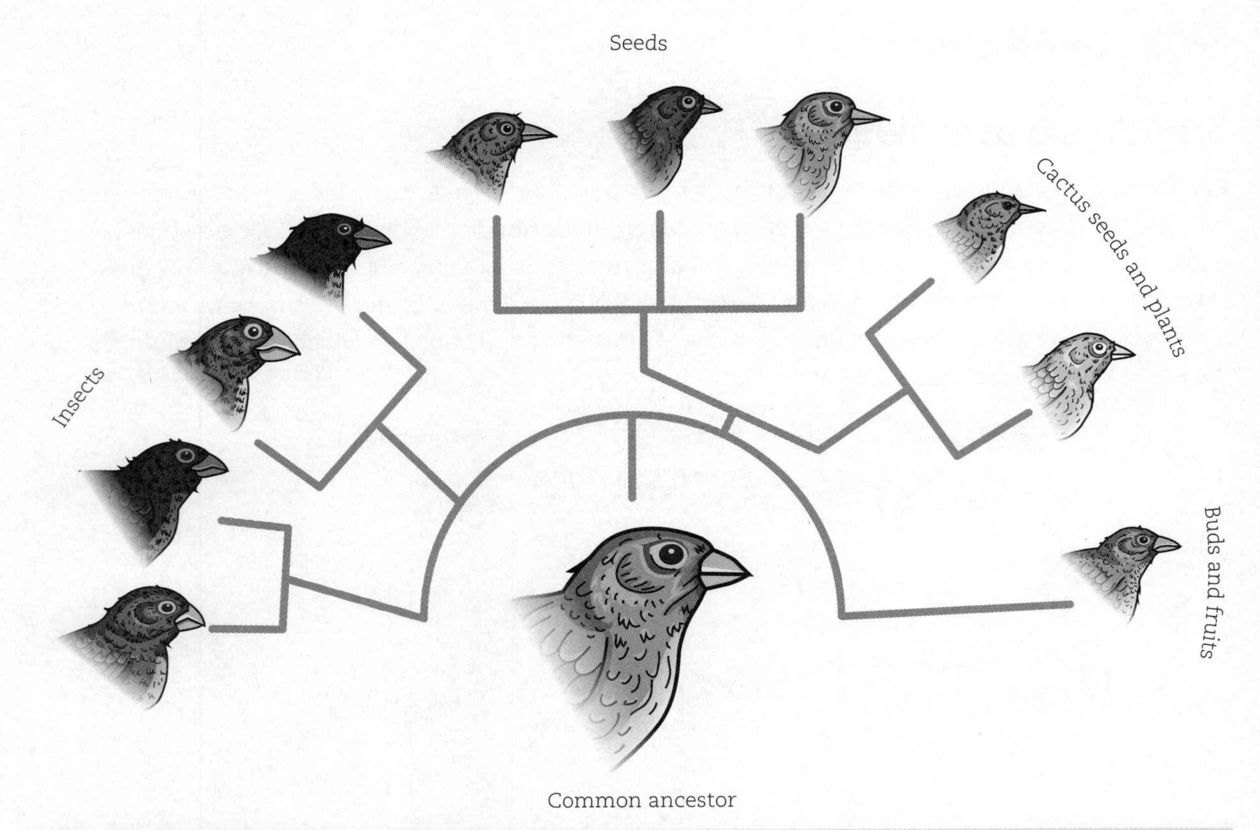

FIGURE 5.2 Adaptive radiation where an ancestor species splits into a number of different populations, likely due to a change in resources or environment. Speciation results as populations become distinct and no longer interbreed.

Hint
Remember: the geological time scale is BIG!

A mass **extinction** occurs when there is change in the environment that results in a decrease in taxonomic diversity. Evidence for this comes from the **fossil** record (Table 5.1). It is estimated that 75–90% of species are lost at a time. Mass extinctions can be followed by a period of evolutionary radiation, as new species emerge to occupy the habitats 'vacated' by the now extinct species.

TABLE 5.1 The five most significant extinctions identified in the fossil record

Million years ago	Era	Loss of
440	Ordovician–Silurian	Small marine organisms
365	Devonian	Tropical marine species
250	Permian–Triassic	Largest number of species on record, including a range of species, including vertebrates
210	Triassic–Jurassic	Vertebrate species (but not dinosaurs)
65	Cretaceous–Paleogene	Non-avian dinosaurs

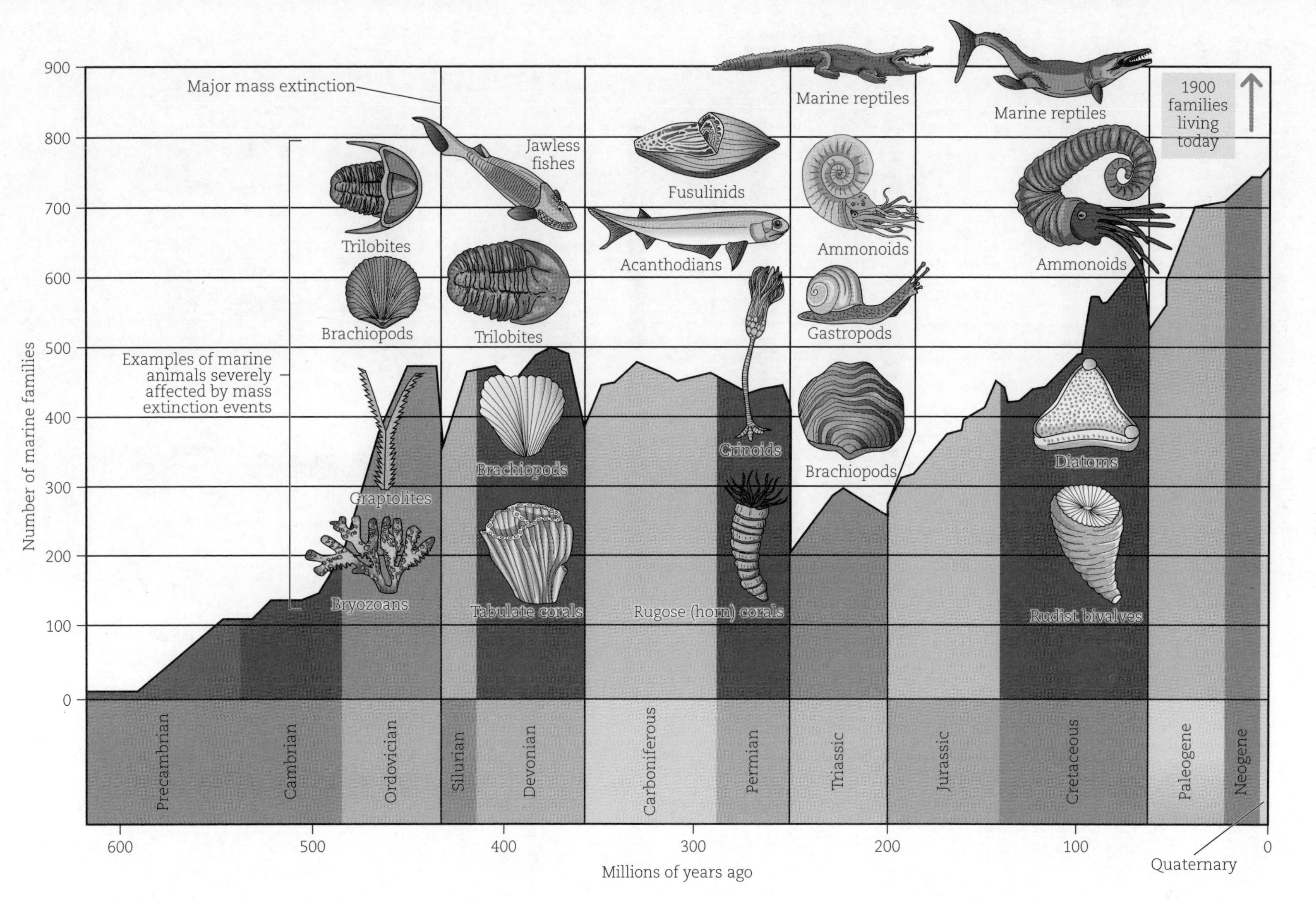

FIGURE 5.3 The fluctuations of marine animal families over the history of the fossil record on Earth

5.1.2 Evidence for the theory of evolution

The original evidence for evolution was based on comparative morphology (in particular, structural homologies), **embryology**, **palaeontology** and **biogeography**.

Developments in technology mean that DNA, RNA and protein sequences are compared to reveal evolutionary relationships. Higher levels of similarity imply a shared molecular history and therefore a common ancestor. The more in common, the more recent the common ancestor.

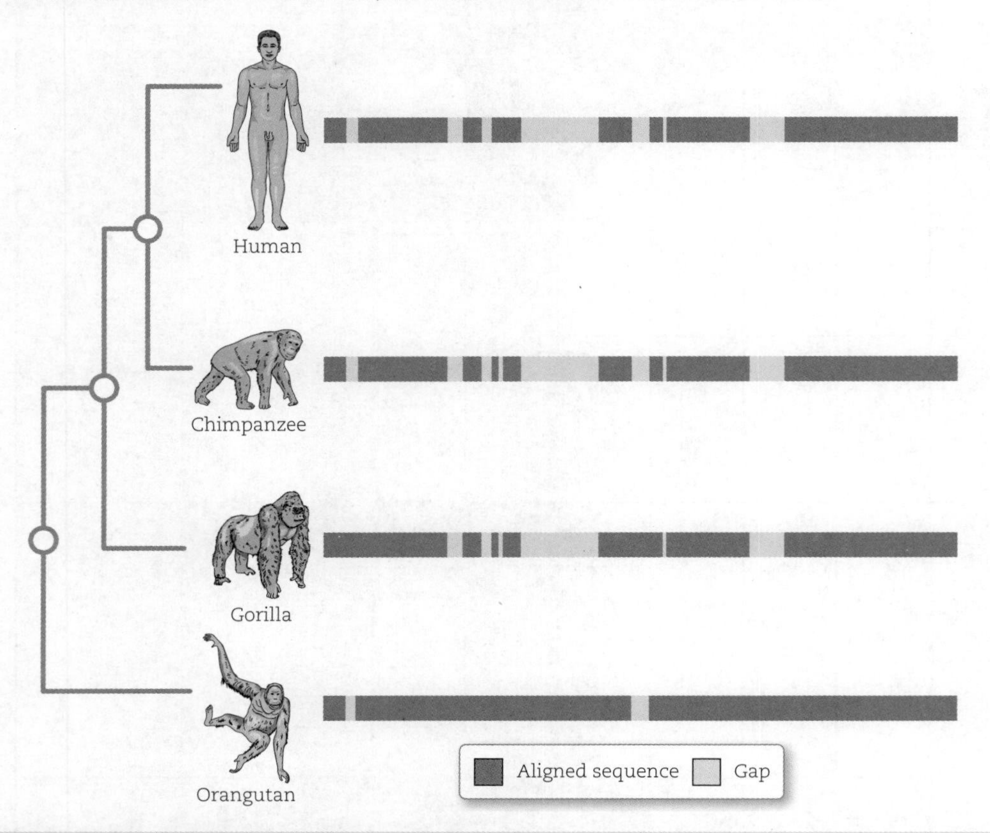

FIGURE 5.4 Comparison of DNA in closely related species

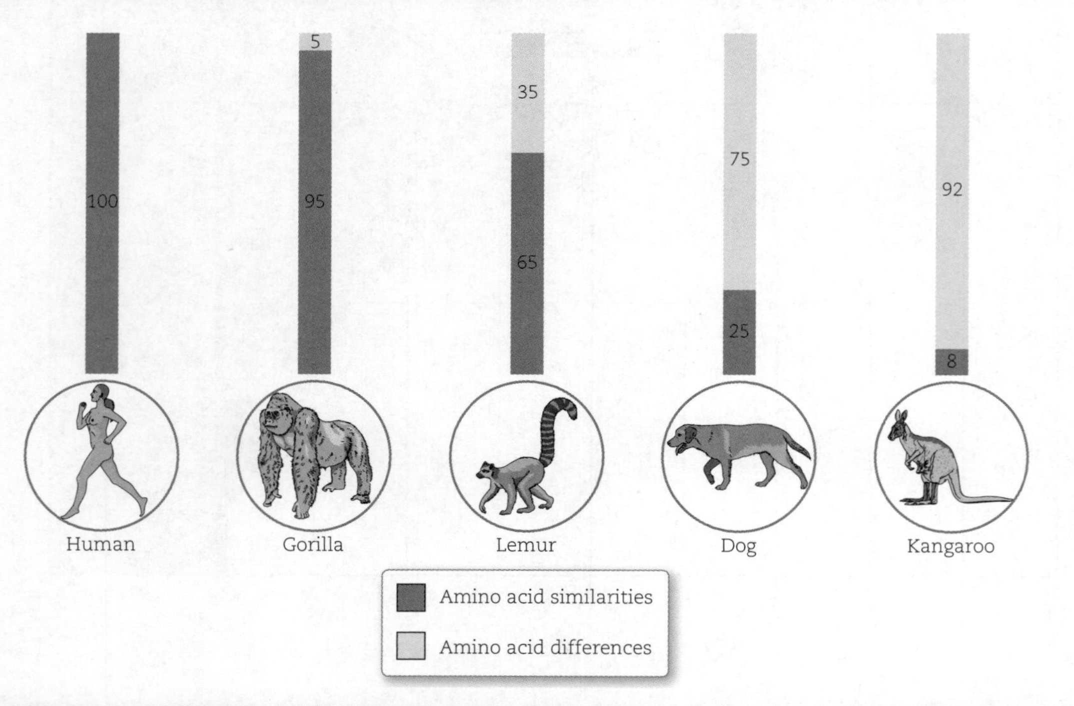

FIGURE 5.5 The percentage of amino acid similarities between some different species for the haemoglobin β-chain

DNA is assumed to accumulate mutations at a relatively constant rate over time and among different species. This means the time since two species diverged from a common ancestor is directly proportional to the genetic difference between them. This is called a molecular clock.

TABLE 5.2 Options for molecular clocks

Sequence being compared	Used for	Reason
Non-coding DNA	Comparing closely related organisms	Mutations occur more readily in non-coding DNA; therefore, they are best for comparisons
Coding DNA	Comparing closely related organisms	Mutations occur more slowly in coding DNA; changes may affect the successful functioning of any proteins produced
Amino acids	Comparing distantly related species	Slowest rate of change because of redundancy in the triplet code

Limitations to molecular clocks include:

- inconsistent mutation rates – either between genes or between species
- initial changes may be masked or reversed by later changes.

Table 5.3 compares an eight-base sequence to illustrate the relationship between molecular data and phylogenetic relationships. The sequences being compared must be homologous (remember 'homologous' from Unit 4 Topic 1).

Species 1 and 2 both have split torsos. They also both have C instead of G in position 3 and T instead of A in positions 7 and 8. These are shared changes from the ancestral group.

Species 3 and 4 both have legs. They also both have G instead of A in position 5. This is one shared change from the ancestral group.

TABLE 5.3 Comparing an eight-base sequence

Species	Sequence							
Ancestral	A	C	G	T	G	T	A	A
1	A	C	**C**	A	G	T	**T**	**T**
2	C	C	**C**	A	G	T	**T**	**T**
3	A	G	G	A	**A**	T	A	T
4	A	C	G	A	**A**	A	A	T

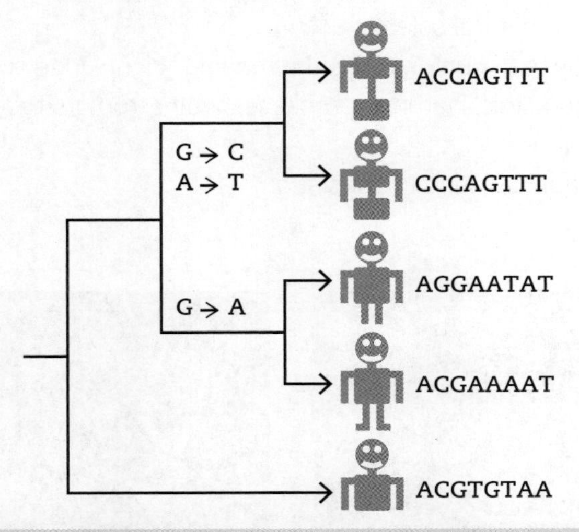

FIGURE 5.6 How to use information from DNA sequencing to inform phylogeny with these hypothetical species. Phylogenetic trees can be based on morphology (presence of legs, presence of split torso) as well as shared sequences. Mutations that are accumulated in the DNA (see Table 5.3) are passed on to descendants.

5.2 Natural selection and microevolution

5.2.1 Natural selection

Natural selection results in the differential survival of organisms over time. It is one of the mechanisms that contributes to evolution (along with mutation, genetic drift and migration).

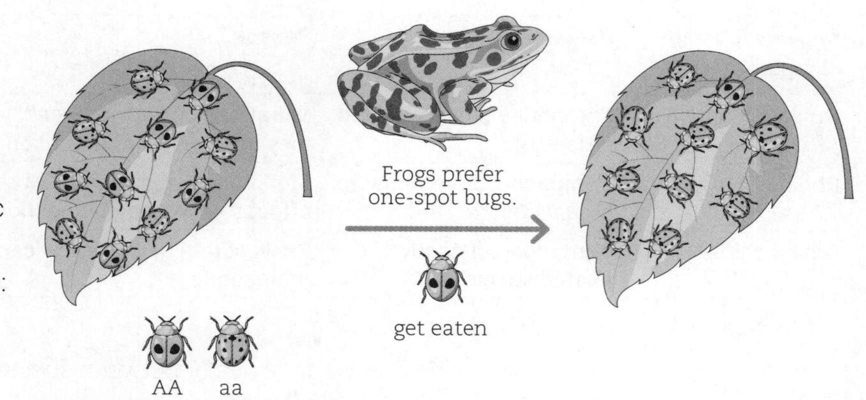

FIGURE 5.7 The effect of a selection pressure on the survival of two phenotypes of ladybug

Natural selection requires:

- traits that are variable in a population and that are heritable
- that not all individuals be able to survive long enough to reproduce, as a result of selection pressures
- that individuals who do survive *and* reproduce pass the genes for their traits on to their offspring.

Therefore, the offspring inherit the trait(s) that increase their own chances of survival and reproduction. This population has **adapted** to its environment.

This can be observed by changes in morphology of populations and by looking at allele frequency for particular traits in a population over time.

Australian examples of adaptation

Hint
Genes for some traits are on the same chromosome as genes that are beneficial to survival and are maintained in the population because they are on the same chromosome, not because they are advantageous.

Australian blind snakes (Figure 5.8) are non-venomous, sightless burrowing snakes that feed on ant eggs and larvae. During embryonic development, eyes are present. However, by the time the snake hatches, it is blind. Natural selection can explain this. Eyes are easily damaged, especially if the animal moves through dirt and is prone to attacks by ants as it is feeding. An animal that hunts its prey by sense of smell without eyes has a better chance of surviving for long enough to reproduce.

The fact that features can be present in embryos shows the gradual nature of the modifications occurring in organisms due to natural selection.

The spinifex hopping mouse (Figure 5.9) has the most efficient kidneys of any mammal. To survive in the arid environment, this mammal processes water from its food and produces highly concentrated, almost solid urine.

Eventually, natural selection can lead to speciation.

Shutterstock.com/Ken Griffiths

FIGURE 5.8 An Australian blind snake

Alamy Stock Photo/Avalon.red/Gerhard Koertner

FIGURE 5.9 A spinifex hopping mouse

5.2.2 Phenotypic selection

Phenotypic selection occurs when more than the expected number of offspring of one phenotype survive compared to any other phenotypes of the trait. Phenotypes of individuals with higher survival rates are described as 'fitter', are more **viable** and have higher **fecundity**.

Factors that affect selection include competition, predation, disease, parasitism (biotic), climate, shelter and catastrophic events (abiotic). The process of selection drives evolution by retaining those individuals (and their alleles) with 'high **fitness**' and removing those individuals (and their alleles) with 'low fitness'. This means selection's effects on allele frequency in a gene pool can be positive or negative. Those organisms that survive are said to have a **selective advantage**.

> **Hint**
> 'Fitness' is a comparative scale. There is no absolute value for fitness.

There are three modes of selection:

1 Stabilising selection – individuals with the intermediate version of the trait have the highest fitness and those at the extremes are selected against. It tends to reduce variation in a population and occurs in stable environments (Figure 5.10).

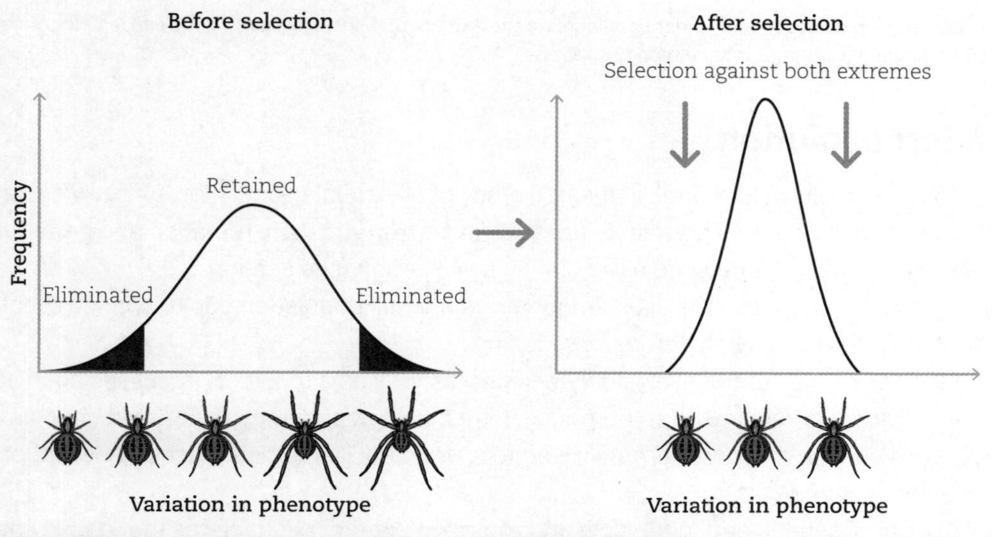

FIGURE 5.10 Stabilising selection in a spider population towards the middle leg length and body size

2 Directional selection – there is a specific direction to the change in the trait towards an extreme. Positive directional selection describes fitness increasing with the trait value. Negative directional selection describes fitness decreasing with the trait value. Directional selection tends to reduce variation in a population and occurs in gradually changing environments (Figure 5.11).

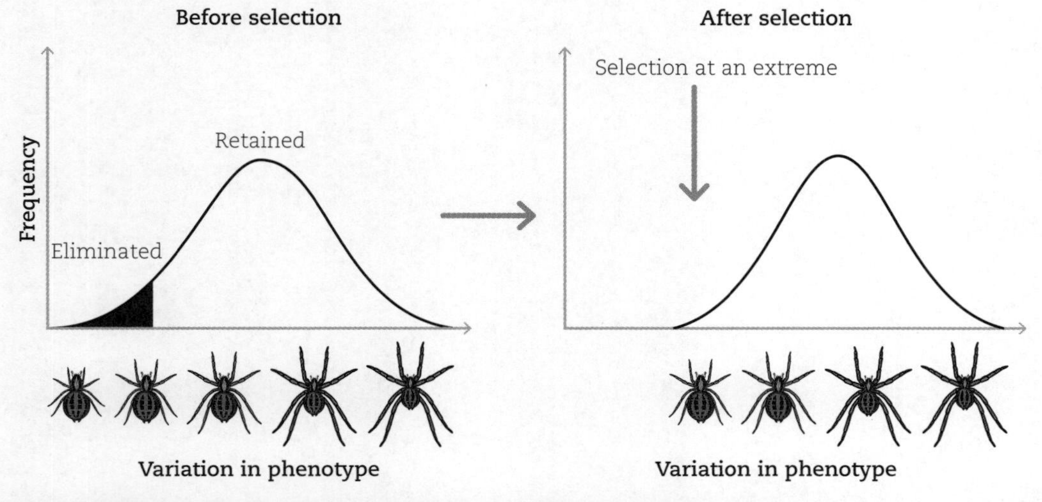

FIGURE 5.11 Directional selection in a spider population towards longer legs and smaller bodies

9780170459136

3 Disruptive selection – individuals with the extreme traits have the highest fitness. Disruptive selection tends to occur in fluctuating environments and may lead to speciation (FIgure 5.12).

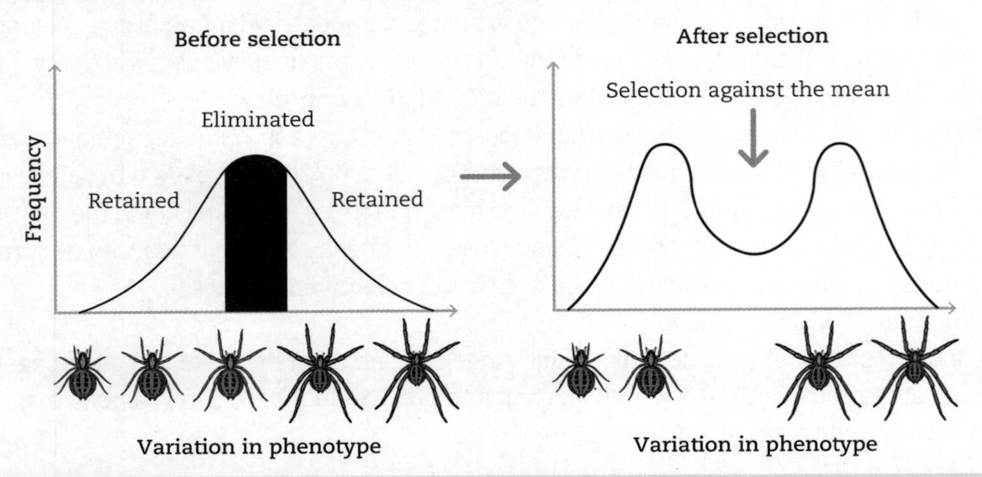

FIGURE 5.12 **Disruptive selection in a spider population to the extremes of short leg/large body** ***and*** **long leg/small body at the expense of medium leg length/ medium body size**

5.2.3 Microevolution

> **Hint**
> Remember: allele frequency is different from genotype and phenotype frequencies.

Microevolution is a small-scale change in the frequency of alleles, in a population, typically occurring over a relatively short period of time. This results from selection, **genetic drift**, mutation and **gene flow**.

New alleles can be introduced into a population by mutation (see section 5.1.2).

Genetic drift is the change in the allele frequency in a small population due to chance events such as emigration or individuals dying before reproducing.

Gene flow is the movement of alleles between populations as a result of processes such as migration of individual organisms that reproduce in their new populations. Restricted gene flow reduces genetic variation and can ultimately lead to speciation. Unrestricted gene flow supports genetic variation in gene pools.

Natural selection, genetic drift, gene flow and mutation are the mechanisms that cause changes in allele frequencies over time.

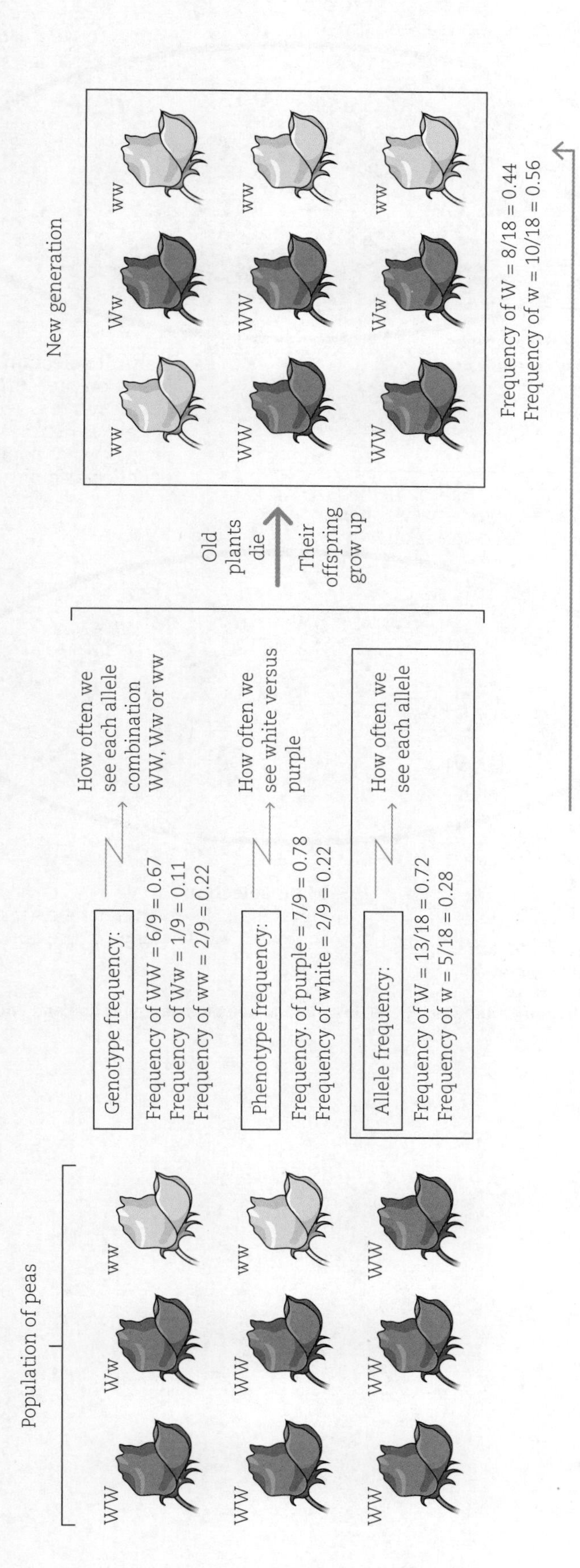

FIGURE 5.13 Changes in allele frequency due to selection pressures. Note: purple flowers are shown in grey.

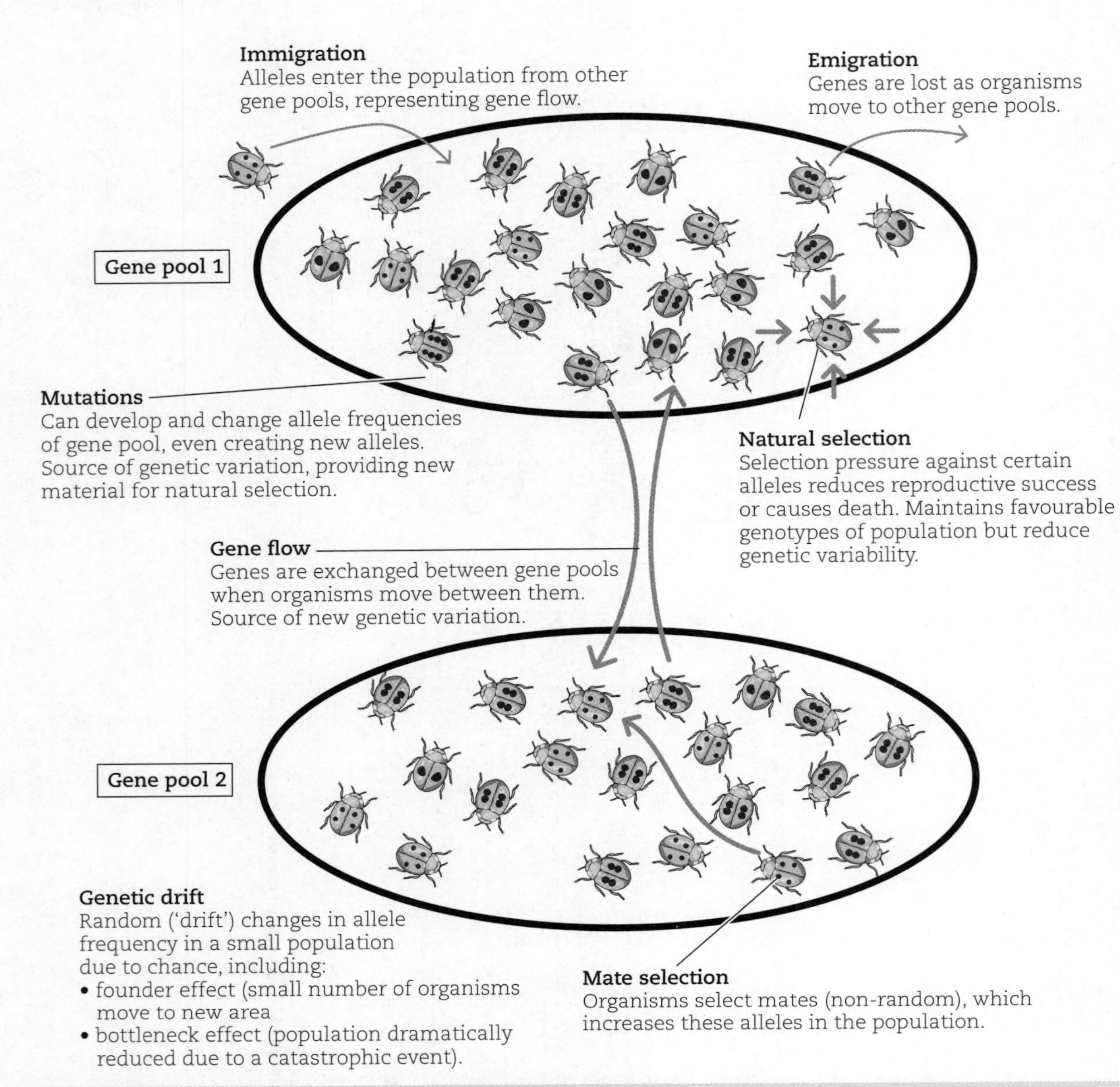

FIGURE 5.14 The effect of genetic drift, mutations, gene flow, natural selection, immigration and emigration on the frequency of alleles in a population

5.3 Speciation and macroevolution

5.3.1 Species diversification

The common patterns of evolution to form a new species that are reproductively isolated are **convergent**, divergent, parallel and coevolution (Table 5.4).

TABLE 5.4 The common patterns of evolution

Evolution pattern	Description	
Convergent	Analogous structures evolve independently in unrelated organisms because they are in similar environments or niches, e.g. flies, kookaburras and fruit bats all have wings.	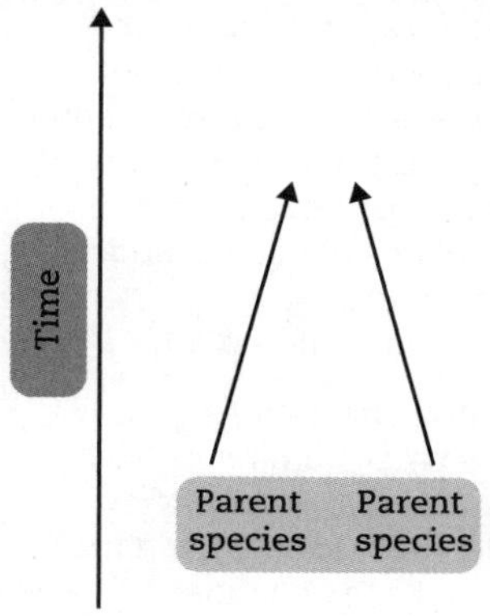 Time; Parent species; Parent species
Divergent	A trait present in a common ancestor persists in different forms in descendants of that ancestor, e.g. the pentadactyl limb in vertebrates (homologous structures).	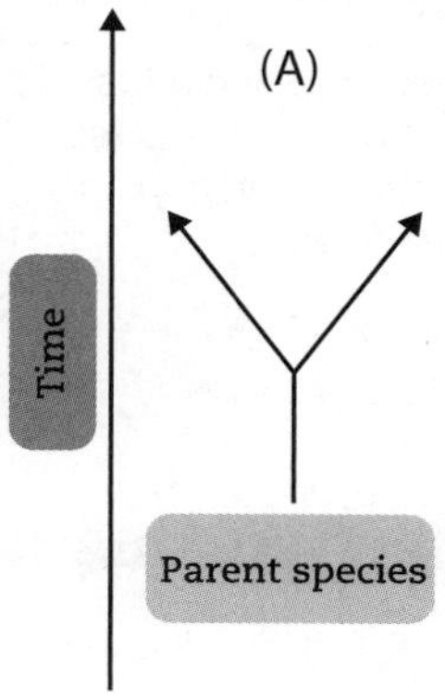 (A); Time; Parent species
Parallel	Similar traits evolve independently of each other in similar environments and the same degree of similarity persists between unrelated species, e.g. marsupial and placental moles.	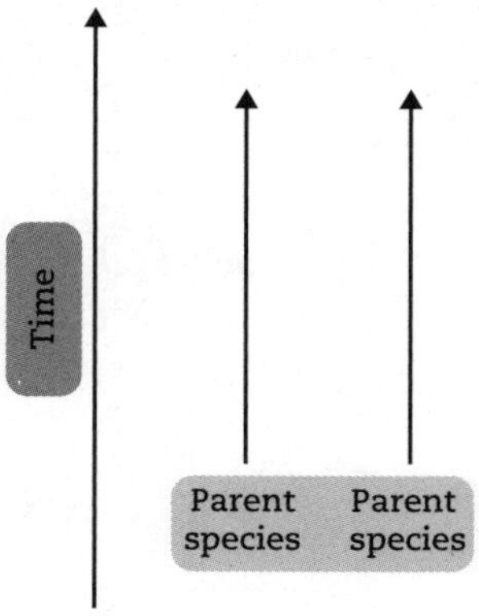 Time; Parent species; Parent species
Coevolution	Often the result of symbiosis or predator–prey interactions and occurs when two or more species exert selection pressures on each other, e.g. ants and acacias.	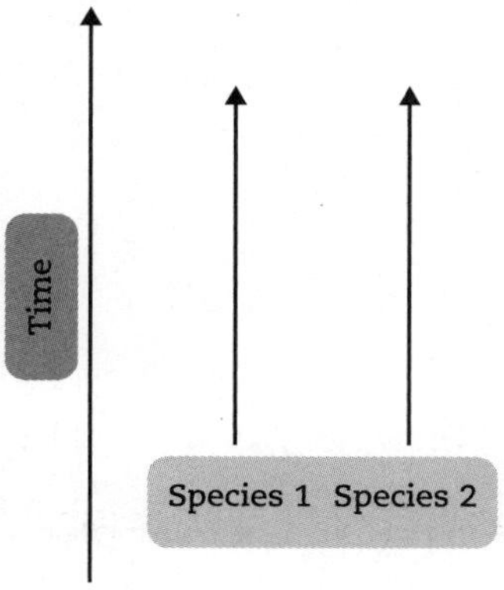 Time; Species 1; Species 2

5.3.2 Modes of speciation

Speciation is the process of forming new species. This occurs when a group within a species acquires unique characteristics after becoming separated from the original population.

Allopatric speciation occurs when populations of the same species become separated geographically and there is a barrier to gene flow.			
Sympatric speciation occurs when populations of the same species are in the same location but, because of reproductive isolation, one population accumulates genetic changes that mean the two populations can no longer interbreed.			
Peripatric speciation occurs when a small number of individuals at the edge of a population's range enter a new niche and can no longer interbreed due to genetic drift.			
Parapatric speciation occurs when populations of a species only have a narrow overlap and minimal gene flow.			

FIGURE 5.15 Speciation

5.3.3 Mechanisms of isolation

Species cannot interact and successfully breed with each other to produce **fertile** offspring because of barriers called isolating mechanisms (see Figure 5.16 and Table 5.5). These mechanisms reduce or prevent gene flow.

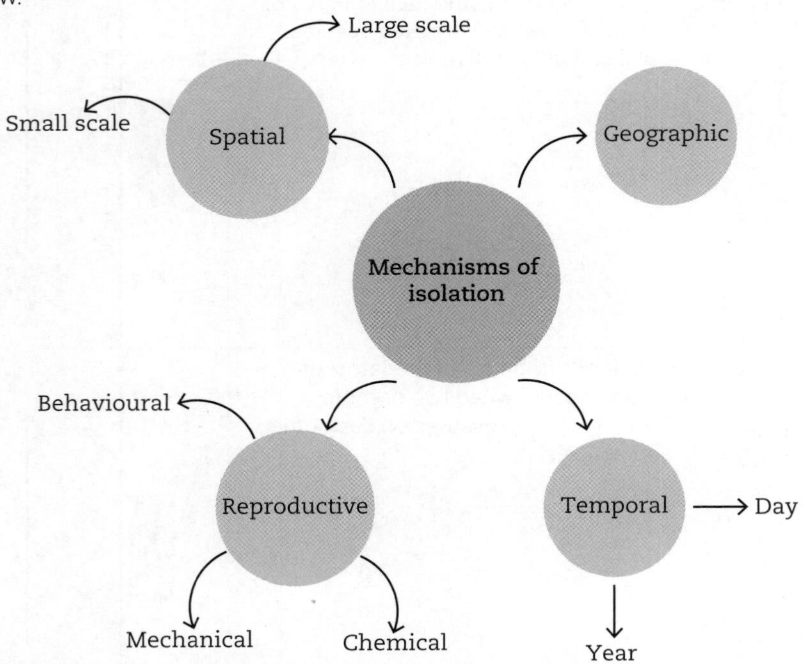

FIGURE 5.16 Summary of isolating mechanisms

TABLE 5.5 Isolating mechanisms

Isolating mechanism	Description	Examples
Geographic	Species cannot come into contact with each other because of a physical barrier, e.g. mountain or river. Habitats can also become fragmented by the action of humans (e.g. clearing for roads) or natural disasters (e.g. cyclone or bushfire).	• Bushfire:  Shutterstock.com/Daria Nipot • Clearing forests:  age-fotostock/IMAGEbroker/Clifton
Spatial	Also described as habitat or ecological isolation and occurs when two populations do not come into contact because of the location of their habitat	• Large scale, e.g. lions and tigers can interbreed, but lions live in grassland of Africa and tigers live in a range of habitats in Asia.  iStock.com/WLDavies Alamy Stock Photo/Image Professionals GmbH/Konrad Wothe • Small scale: This can also occur within habitats between niches, e.g. non-biting midges of the *Chironomus* species. 

››

››

Isolating mechanism	Description	Examples
Reproductive	Includes behaviours, structures and physiological differences to prevent fertilisation	• Mating or courtship behaviours, e.g. *Maratus* spiders dance and vibrate to attract a mate, with each species having a specific series of movements.  Alamy Stock Photo/Paul Harrison • Mechanical isolation means that reproductive structures do not fit together, e.g. orchid and bee. Many native terrestrial orchids mimic nectar-bearing flowers for the purpose of pollination. The tiger orchid (*Diuris sulphurea*) replicates the structure of bush peas and emits a pheromone similar to that of a female bee. This attracts male bees, who attempt to mate with the flower, pollinating the orchid in the process.  Shutterstock.com/Anne Powell • Chemical isolation that only allows sperm or pollen from the correct species to fertilise eggs, e.g. spawning animals that release eggs and sperm into the water. Incompatibility genes, which prevent pollen from germinating or growing into the stigma of a flower, have been discovered in many angiosperm species. If plants do not have compatible genes, the pollen tube stops growing. Self-incompatibility is controlled by the S (sterility) locus. 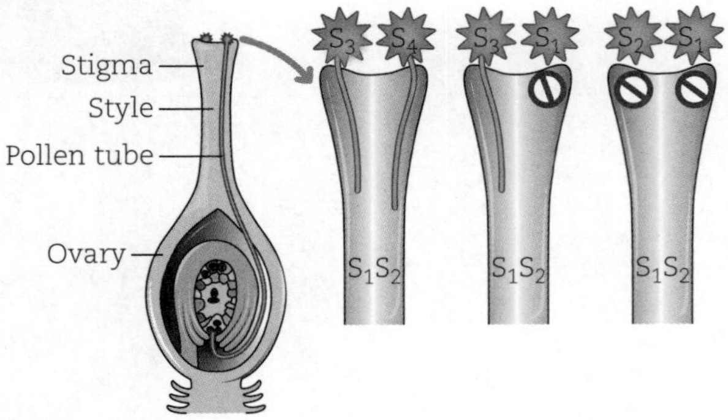 Stigma Style Pollen tube Ovary S_3 S_4 S_2 S_1 S_2 S_1 S_1S_2 S_1S_2 S_1S_2

››

Isolating mechanism	Description	Examples
Temporal	Occurs when species become reproductively active at different times of the day or seasons of the year	• E.g. *Acropora* populations at Scott Reef spawning at different times, autumn and spring. Nature Picture Library/Jurgen Freunda • The adults of two closely related species of cicadas of the genus *Magicicada* can breed with each other. However, *M. tredecim* emerge every 13 years, whereas those of *M. septendecim* only emerge every 17 years. Species may also reach reproductive maturity at different ages. Alamy Stock Photo/Larry Postell

These mechanisms all prevent the fertilisation of an egg and formation of a zygote. Therefore, they are considered to be pre-zygotic isolating mechanisms.

Post-zygotic mechanisms include zygote mortality, hybrid inviability, hybrid sterility, hybrid breakdown.

5.3.4 Extinction risks

An understanding of how natural selection, genetic drift, gene flow and mutation work together is important if we want to conserve populations and reduce the risk of extinction. All populations are subject to the effect of genetic drift. The smaller the population, the greater the risk to genetic diversity and population viability.

Population bottlenecks, caused by hunting, natural disasters and habitat destruction, significantly reduce the size of a population and its gene pool for at least one generation, resulting in low genetic diversity. Hence, random fluctuations in alleles (genetic drift) have an increased effect. Low-frequency alleles have a greater chance of being lost, further reducing the gene pool. This means that the population may not be able to respond to selection pressures because the allele required may have 'drifted' out of the population. This may lead to extinction *or* speciation or the population may recover.

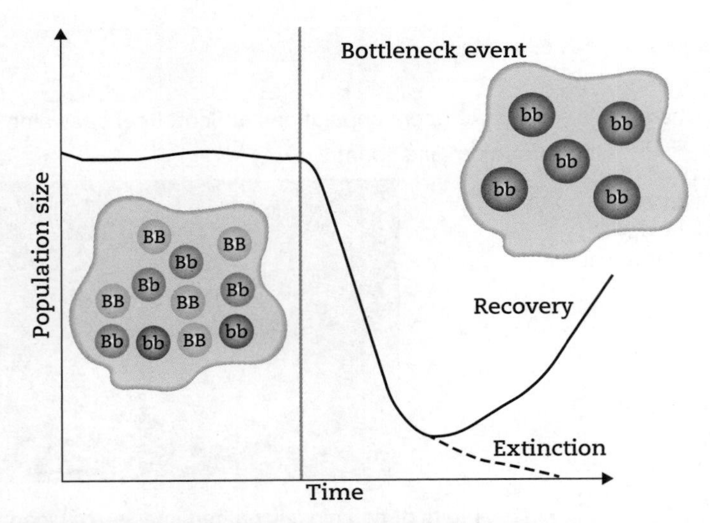

FIGURE 5.17 The effect of a bottleneck on population size and allele frequency

An Australian example of a population bottleneck is seen in black-footed rock wallabies (Figure 5.18). Black-footed rock wallaby distribution has dropped 93% in South Australia. Reintroduced populations have lost significant genetic diversity.

FIGURE 5.18 Black-footed rock wallabies suffered a population bottleneck.

Founder effects are similar to bottlenecks but occur when a small number of individuals from an original population separate to form a new population (Figure 5.19). This small group of individuals may contain a non-random sample of the alleles in the population. The resulting reduced genetic variation compared to the original population exposes them to the same risks as a population after a bottleneck.

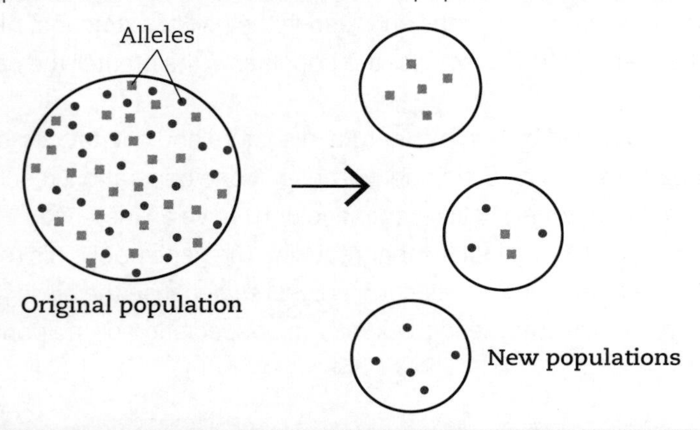

FIGURE 5.19 Founder effects showing new populations with reduced allele variation compared to the original population (alleles represented by different shapes)

FIGURE 5.20 The population of koalas on Kangaroo Island is an example of the founder effect.

An Australian example of the founder effect is seen in koalas. Many male koalas from Kangaroo Island only have one testicle. A lack of genetic diversity has led to testicular aplasia, although this has not affected their reproductive ability. It probably resulted from the selection of founder individuals carrying alleles for testicular abnormalities. These alleles then increased through genetic drift.

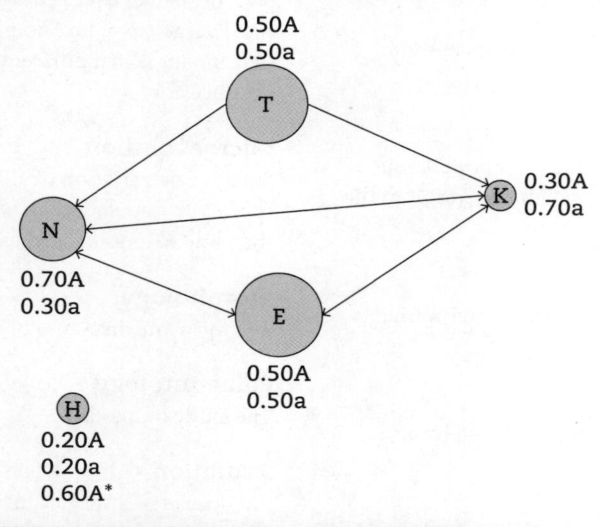

FIGURE 5.21 Arrows showing gene flow between populations E, H, K, N and T of one species

The syllabus requires analysis of gene flow and allele frequency data.

Figure 5.21 shows gene flow between populations of one species. Note that gene flow is into populations N, E and K. Gene flow is out of population T. Population H is isolated. Because of their isolation, populations T and H are most likely to be undergoing speciation. Population H may be an example of allopatric speciation where a geographic barrier has arisen (there is no gene flow at all) or peripatric speciation, because population H may be occupying a new niche.

Allele frequencies can also show the outcome of gene flow. Even though the allele frequencies of A and a are different between each population, population H is the most likely to undergo speciation because there is no gene flow and it has acquired the mutant allele A*.

Glossary

A+ DIGITAL FLASHCARDS
Revise this topic's key terms and concepts by scanning the QR code or typing the URL into your browser.

https://get.ga/aplus-qce-bio-u34

Adapted
A change in structure or function that is inherited that results in an increase in ability to survive and reproduce.

Biogeography
The distribution of a species through space and time.

Convergent
A type of evolution of similar traits in response to similar environmental factors in different species.

Divergence
When two populations of a single species accumulate enough differences through mutation to be reproductively isolated.

Embryology
The study of foetal development from conception to birth.

Evolution
Change in the genetic composition of a population during successive generations, which may result in the development of a new species.

Extinction
The disappearance of a species from all environments on Earth.

Fecundity
Fertility; the number of offspring produced by an individual.

Fertile
Capable of producing viable offspring.

Fitness
A measure of reproductive and adaptive success.

Fossil
The petrified remains or impressions of a once-living organism.

Gene flow
The transfer of genetic material, alleles, from one population to another.

Genetic drift
The change in the frequency of a gene in a population due to the random processes of reproduction.

Isolation
When two populations, who could interbreed, become separated spatially, temporally, geographically or behaviourally.

Macroevolution
The variation of allele frequencies at or above the level of species over geological time, resulting in the divergence of taxonomic groups, in which the descendant is in a different taxonomic group from the ancestor.

Microevolution
Small-scale variation of allele frequencies within a species or population, in which the descendant is of the same taxonomic group as the ancestor.

Morphology
The form and structure of an organism.

Palaeontology
The study of fossils.

Radiation
The spread of a species throughout a habitat.

Spatial
Relating to space.

Selective advantage
The traits or characteristics that are heritable and increase the likelihood of an organism surviving and reproducing in a particular environment.

Temporal
Relating to time.

Viable
Can live, grow and reproduce.

Revision summary

Use the following checklist of syllabus dot points and key knowledge within Unit 4 Topic 2 to ensure that you have thoroughly reviewed the content. Provide a brief definition or comment for each item to demonstrate your understanding or code them using the traffic light system – green (all good), amber (needs some review), red (priority area to review).

Evolution	
• define the terms *evolution*, *microevolution* and *macroevolution*	
• determine episodes of evolutionary radiation and mass extinctions from an evolutionary time scale of life on Earth (approximately 3.5 billion years)	
• interpret data (i.e. degree of DNA similarity) to reveal phylogenetic relationships with an understanding that comparative genomics involves the comparison of genomic features to provide evidence for the theory of evolution.	
Natural selection and microevolution	
• recognise natural selection occurs when the pressures of environmental selection confer a selective advantage on a specific phenotype to enhance its survival (viability) and reproduction (fecundity)	
• identify that the selection of allele frequency in a gene pool can be positive or negative	
• interpret data and describe the three main types of phenotypic selection: stabilising, directional and disruptive	

››

›› • explain microevolutionary change through the main processes of mutation, gene flow and genetic drift.	
• Mandatory practical: Analyse genotypic changes for a selective pressure in a gene pool (modelling can be based on laboratory work or computer simulation).	
Speciation and macroevolution	
• recall that speciation and macroevolutionary changes result from an accumulation of microevolutionary changes over time	
• identify that diversification between species can follow one of four patterns: divergent, convergent, parallel and coevolution	
• describe the modes of speciation: allopatric, sympatric, parapatric	
• understand that the different mechanisms of isolation — geographic (including environmental disasters, habitat fragmentation), reproductive, spatial, and temporal — influence gene flow	
• explain how populations with reduced genetic diversity (i.e. those affected by population bottlenecks) face an increased risk of extinction	
• interpret gene flow and allele frequency data from different populations in order to determine speciation.	

Exam practice

Multiple-choice questions

Each multiple-choice question is worth 1 mark.

Solutions start on page 164.

Question 1

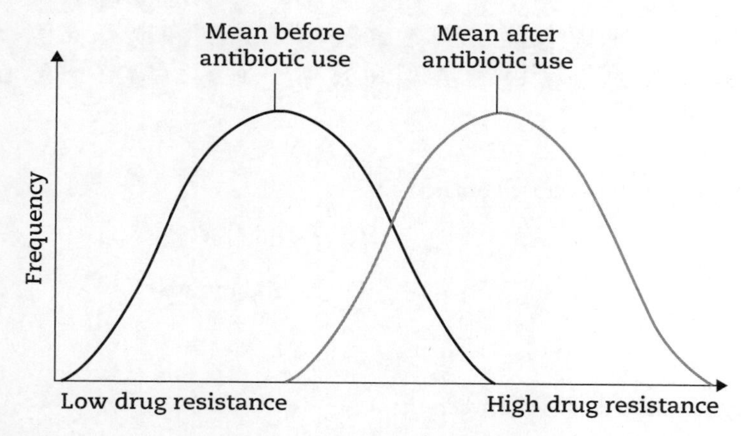

The graph above shows the frequency at which a bacterial species was detected in soil before and after antibiotic use. The shift in the mean is best described as

A destabilising selection.

B directional selection.

C disruptive selection.

D stabilising selection.

Question 2

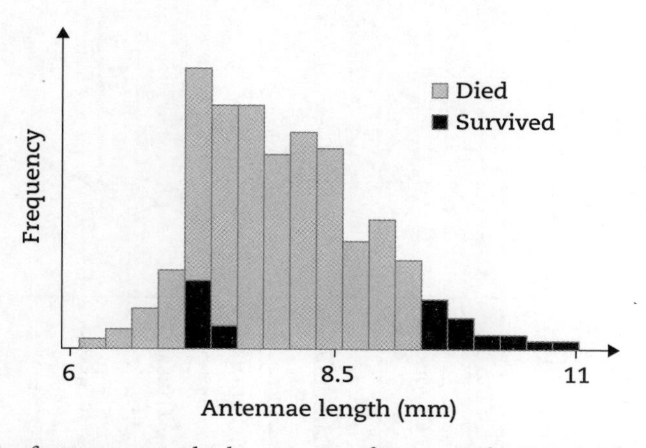

The graph above shows the frequency at which a species of insect with antennae length ranging from 6 mm to 11 mm occurred before and after a season of drought. The shift in the population is best described as

A destabilising selection.

B directional selection.

C disruptive selection.

D stabilising selection.

The following information relates to Questions 3 and 4.

The diagram below shows the same protein sequence in several different species. Each letter represents an amino acid.

Mouse	F	S	T	A	A	F	R	F	G	H	A	T	V	H	P	L	V	R	R	L	N	T
Rat	F	S	T	A	A	F	R	F	G	H	A	T	V	H	P	L	V	R	R	L	N	T
Human	F	S	T	A	A	F	R	F	G	H	A	T	I	H	P	L	V	R	R	L	D	A
Pig	F	S	T	A	A	F	R	F	G	H	A	T	I	H	P	L	V	R	R	L	D	A
Dog	F	S	T	A	A	F	R	F	G	H	A	T	V	H	P	L	V	R	R	L	D	A
Chicken	F	A	T	A	A	F	R	F	G	H	A	T	I	Q	P	L	V	R	R	L	N	A
Frog	F	T	T	A	A	F	R	F	G	H	A	T	I	P	P	M	V	H	R	L	D	S

Question 3

Identify which animals are the most closely related.

A dog and frog

B frog and pig

C pig and rat

D rat and mouse

Question 4

The biggest difference in amino acids is between the

A dog and frog.

B frog and pig.

C pig and rat.

D rat and mouse.

Question 5 ©QCAA QCAA 2020 P1 MC Q5

The following graph shows the diversity of marine animals since the late Precambrian. The data is from marine animal families that have been reliably preserved in the fossil record.

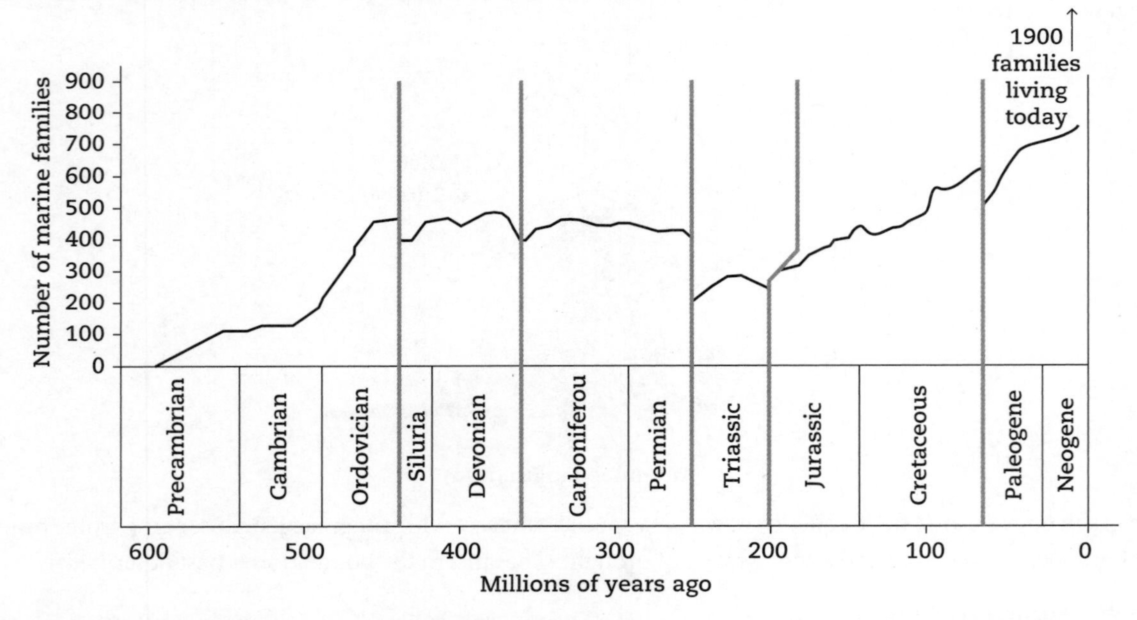

Which of the following time periods had the greatest evolutionary radiation of the marine families?

A Ordovician

B Cretaceous

C Devonian

D Permian

Question 6 ©QCAA QCAA 2020 P1 MC Q20

Which of the following are features of both microevolution and macroevolution?

A mutations only

B mutations and gene flow only

C gene flow and genetic drift only

D mutations, gene flow and genetic drift

Short-response questions

Solutions start on page 165.

Question 7 (1 mark)

Define 'evolution'.

Question 8 (3 marks)

A short region of a DNA sequence for a gene from four different organisms is shown below.

Organism W: TGG GCT AAC AAG CAA ATG ATC T

Organism X: TGG GCT ACC AAG CAA ATG ATC T

Organism Y: TGG GCT AAC AAG CAA ATG ATC T

Organism Z: TCC CCT ATC AAG GAA ATG ATA T

Interpret the sequences to describe the phylogenetic relationship between these four organisms.

Question 9 (2 marks)

The diagram below shows a population of ladybugs over time.

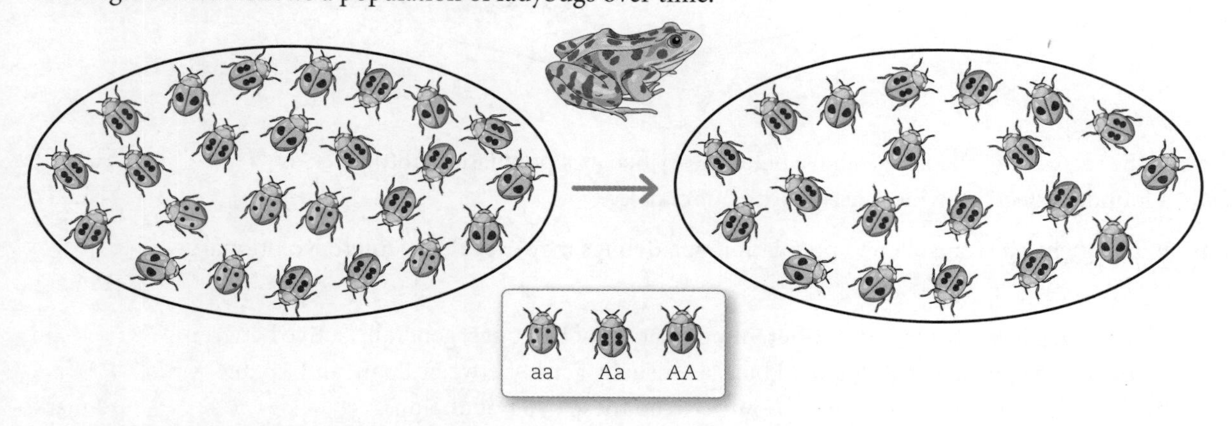

a What is the selection pressure in the scenario above? 1 mark

b Describe the selection for the a allele as positive or negative in this population. 1 mark

Question 10 (6 marks)

Some ladybugs were accidentally removed from a bushland habitat to an urban garden when a shrub was transplanted.

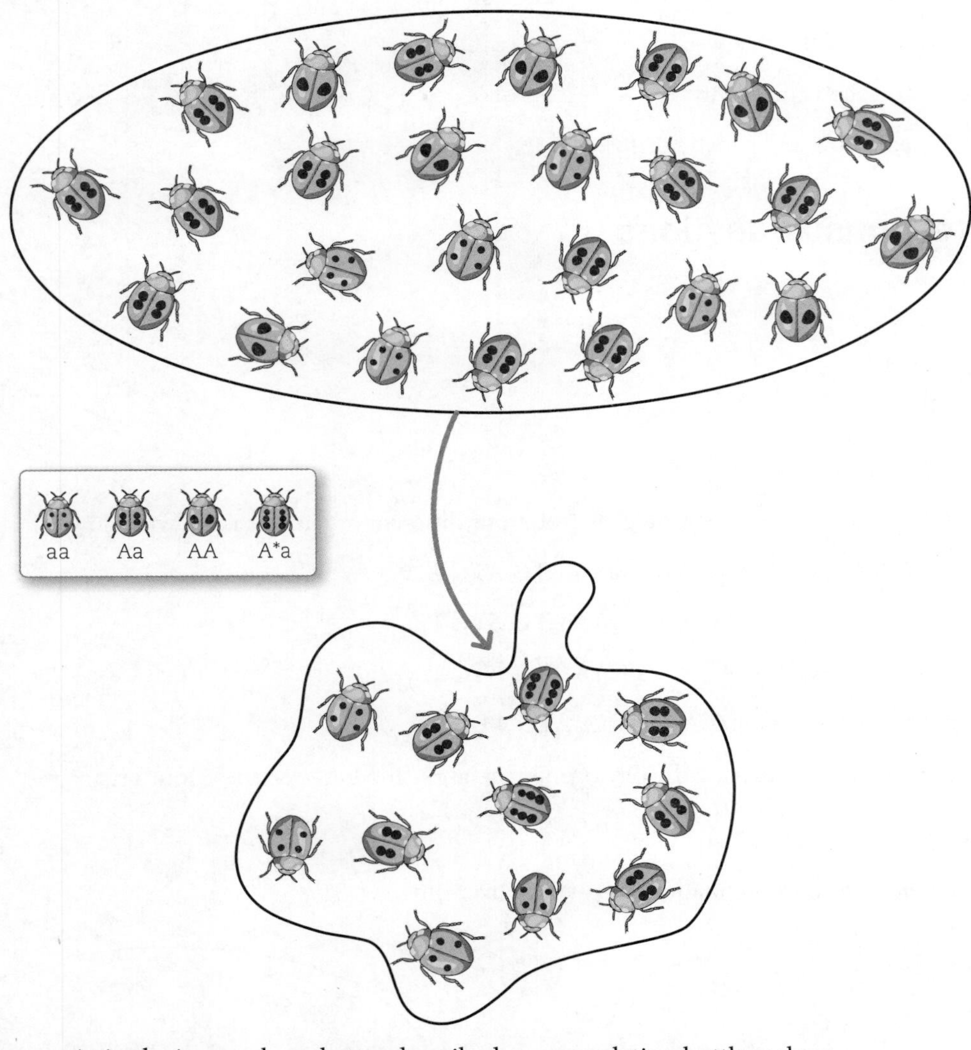

a Is the scenario in the image above better described as a population bottleneck or a founder effect? Provide a reason for your choice. 2 marks

b Explain why the transplanted population of ladybugs may experience microevolutionary changes. 2 marks

c The urban garden contained a different population of ladybugs, generally active between 5 a.m. and 10 a.m. The transplanted bugs are usually active between 1 p.m. and 5 p.m. Explain if there is likely to be gene flow between these two populations. 2 marks

Question 11 (8 marks)

The following diagram represents tiger populations in three national park areas in Nepal. Around each national park, forests have been cleared for agriculture and timber, as well as the building of road networks and other development activities.

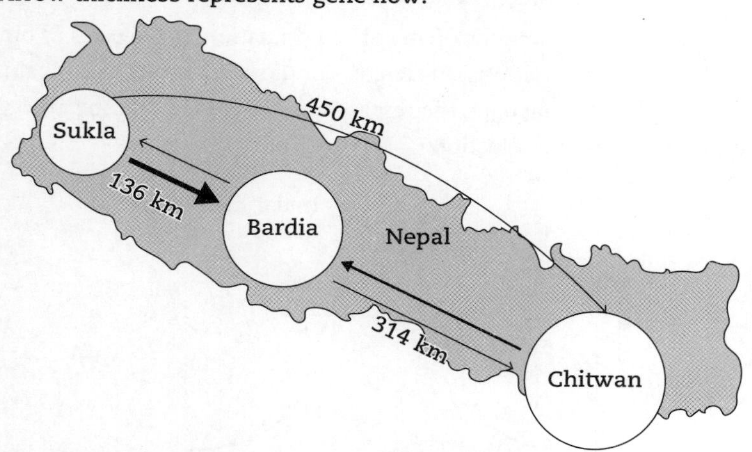

a Use information in the diagram to describe the most likely type of speciation the tigers of the Sukla population will undergo. 2 marks

b Explain why the population of tigers at Sukla is at greater risk of extinction than those at Bardia or Chitwan. 4 marks

c Identify evidence of habitat fragmentation. 1 mark

d Provide a possible explanation as to why the gene flow between the Chitwan and Sukla populations is different from the gene flows between the Sukla and Bardia populations and between the Chitwan and Bardia populations. 1 mark

Question 12 (2 marks)

Scenario 1:

Many species of bats provide dispersal and pollination services to a large number of plants species as a result of their diet. *Crescentia alata* is a tree native to southern Mexico and Costa Rica. The flowers of this tree open at night. *Crescentia alata* is unable to reproduce without bats acting as pollinators.

Scenario 2:

Australia has the world's greatest variety of marsupials. Many marsupials are similar in size and shape to placental mammals on other continents, including homologous structures such as the pentadactyl limb. This suggests that many species of mammals descended from the same common ancestor.

There are four identified patterns of evolution: divergent, convergent, parallel and coevolution.

Identify the type of evolution best represented by scenarios 1 and 2.

Question 13 (3 marks) ©QCAA QCAA 2020 P2 SR Q9

Fossil evidence seems to show that the morphology of the Queensland lungfish has remained relatively unchanged for the past 100 million years. Describe the features of the theory of natural selection to explain how this may have occurred.

Question 14 (3 marks) ©QCAA QCAA 2020 P2 SR Q10

Researchers measured the adult beak lengths of an entire population of a species of bird and plotted their results on the graph. After many generations, the lengths of the adult beaks were again measured. By comparing this new data to the original data, the researchers concluded that the average length of beaks had increased as a result of directional selection.

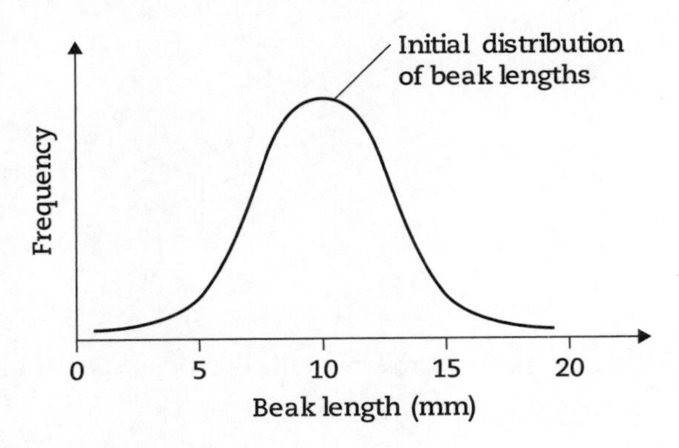

a Describe directional selection. 2 marks

b On the graph provided, sketch a representation of directional selection for the beak length scenario. 1 mark

Question 15 (6 marks) ©QCAA QCAA 2020 P2 SR Q11

The image shows changes in the frequency of a particular gene in a single species of bird, leading to a speciation event. These changes have occurred over a period of successive time points (i.e. I, II and III) each separated by approximately 1000 generations.

- The letters A, B, C and D represent separate niches inhabited by the birds.
- The arrows depict gene flow between the niches.
- The allelic frequency for the gene is shown as *f* in each niche.

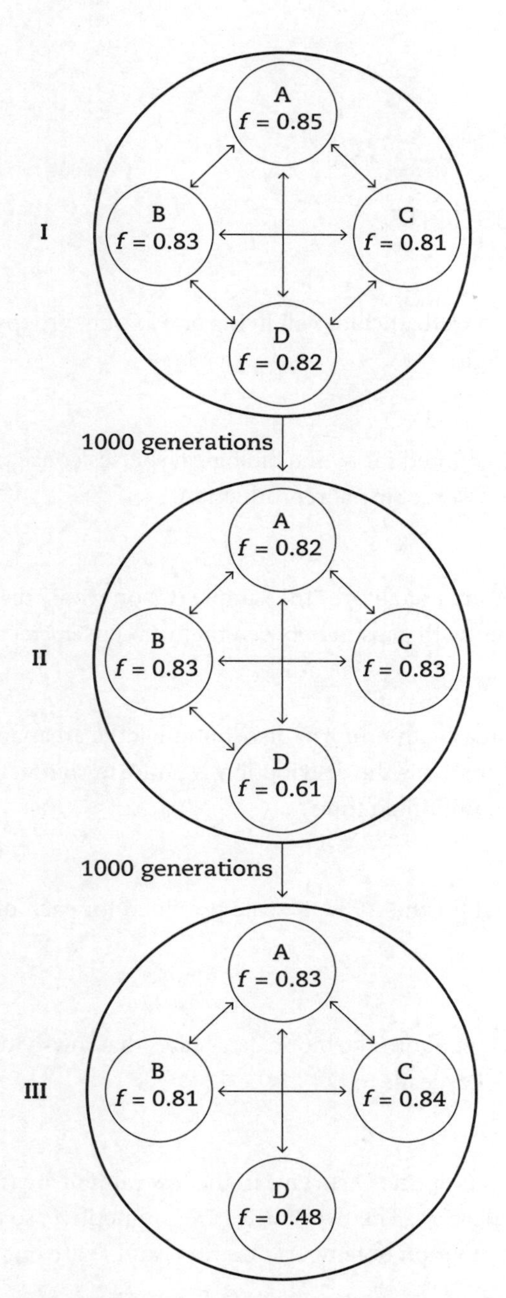

Draw a conclusion about the type of speciation that has occurred in this population. Explain your reasoning by referring to the information provided in each of the time points.

SOLUTIONS

CHAPTER 1: UNIT 3 TOPIC 1

Multiple-choice questions

Question 1 C

C is the correct answer because a clade includes all living and extinct groups (not D) of organisms (not A or B) with a common origin.

Question 2 C

Responses A, B and D are all accounted for by the biological species concept. C is not because the biological species concept does assume sexual reproduction.

Question 3 B

There are dots (not A), the dots are not all over the sample (C) or evenly distributed (D). However, the dots are grouped closely together with gaps in between them, so B is correct.

Question 4 B

Random sampling is used for areas with uniform distribution, large areas and limited time. Therefore, not A because there is a gradient, C because the distribution is uniform so don't need further information, or D because the area is small with unlimited time.

Question 5 D

D matches the percentage stated for the ACFOR scale provided for each of the lichens, alga 1 and alga 2.

Question 6 D

Soil is acidic (pH < 7) and nutrient poor. The biotic data states that low-lying grasses (not moss or lichen) are predominant. Following the key leads to D.

Question 7 B

The key point is low rainfall. B lets Species I survive in the low rainfall areas, while Species II out-competes it in the higher rainfall areas. The two species are competitive, so the distribution must change, so A is not correct. There is no mention of temperature, so C and D are not relevant.

Short-response questions

Question 8

a *Melaleuca* low woodland (1 mark)

If the name of the dominant species is given, it is used as part of the name.

b The dominant species of the canopy is the eucalypt (1 mark). The height of the trees is 10–30 m. (1 mark) The canopy cover is 31–50%. (1 mark)

c There are **two clearly different areas or strata** in the habitat being sampled. (1 mark)

d The ecologist needed to know the percentage cover of the canopy, which cannot be determined by transects. (1 mark)

The ecologist did not need to determine the distribution of plants along a gradient; therefore, quadrats are more appropriate. (1 mark)

Make sure your answer refers to both quadrats and transects, since both are in the question.

e Any two of the following for 1 mark each:

- Increasing the size of the sample so that data better reflects the habitat
- Choosing an appropriate size of quadrat to measure percentage cover
- Consistent counting criteria to determine percentage cover in a quadrat
- Ensuring that a proportionate number of samples is taken from each of the two strata/areas
- Noting the precision of the equipment – quadrats are less precise than densiometers

Make sure you link your answer to the data provided.

Do not accept: random number generator – there is nothing in the data provided that says the sampling is random.

Do not accept: calibrating equipment properly – it is not necessary to calibrate a quadrat.

Question 9

a 1 mark for either of the following:

- a circle that includes the brown and Peruvian pelican with the common ancestor node.
- a circle that includes the brown, Peruvian and American pelicans that includes the common ancestor node.

b An 'X' on the first node. (1 mark)

c

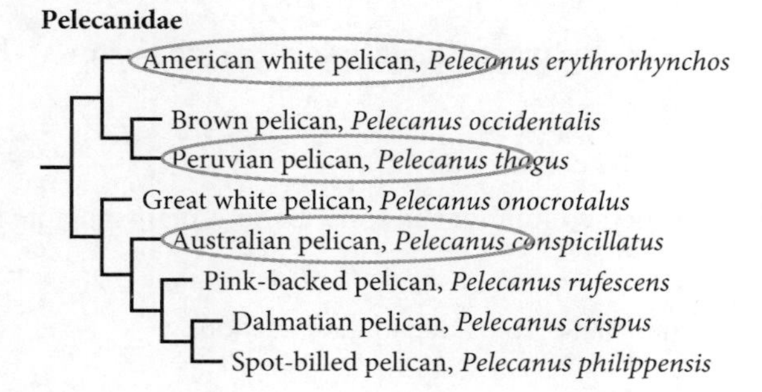

American white pelican

Remember: this shows genetic distance, and the distance from the common ancestor is shorter to the American white pelican than to the Australian pelican.

Question 10

a 1 mark for correct substitution.

Show correct substitution into the formula in some way.

$$SDI = 1 - \frac{7(6) + 15(14) + 20(19) + 6(5)}{48(47)}$$

OR

$$SDI = 1 - \frac{662}{2256}$$

OR

SDI = 1 – 0.29

Remember to subtract your answer from 1.

SDI = 0.71 1 mark for the correct answer and it *must* be to 2 decimal places. (Answers with incorrect rounding and/or 1 or 3 decimal places will be marked incorrect.)

b Similarity: both sites have these species: grey mangrove, algae fish. (1 mark)

Difference: site 2 does not have the pelican. (1 mark)

Significance: this means that site 1 has a higher species richness because there are more types of species present. (1 mark)

c Salinity is a factor that may limit abundance because Graph 1 shows a lower growth rate at higher salinities. (1 mark)

Light intensity does not appear to limit abundance because the height of seedlings is consistent across a number of different light intensities. (1 mark)

Both sources stated in the question are referred to in the answer.

Question 11

a Species 3 as it is most likely to be r-selected with a **high reproduction rate** (1 mark) that is typical of asexually reproducing species, but a **lower survival rate** due to low/non-existent parental care. (1 mark)

b Species 2 (1 mark)

Any two of the following reasons (1 mark each):

- Have only one offspring at a time
- High parental care of 1–2 years
- Higher survival rate of offspring.
- Reproductive age is late relative to life span of 12 years.

c K-selected (1 mark)

Question 12

a Invertebrates → fish (1 mark)

Answer must show that the fish prey on invertebrates generally *or* a specific, named invertebrate.

b Mutualism (1 mark if with the correct pairing listed below)

Crabs aerate the soil – benefits plants; plants stabilise soil for crab burrows – benefits crabs. (1 mark)

OR

Commensalism (1 mark if with the correct pairing)

Invertebrates increase the nitrogen content of the soil – benefit plants; plants neither benefit/harm the invertebrates. (1 mark)

You must base your evidence on information provided in the question.

c Interspecific competition (1 mark)

It is out-competing different species, not other *Spartina* plants. (1 mark)

Question 13

Water stress (1 mark)

In example A, there was 90% survival with low water stress. Example C had 19% survival with high water stress. (1 mark)

Both examples A and C had cold temperatures and ground cover protection; the main difference was water stress. (1 mark)

Shade (1 mark)

In example B, there was 44% survival with full sunlight. Example D had 62% survival with shade. (1 mark)

Examples B and D had the same conditions; the main difference was shade OR sunlight. (1 mark)

Question 14

a A clade is a group of organisms that consists of a common ancestor and all its lineal descendants. One clade is shown below; there is more than one correct answer. (1 mark)

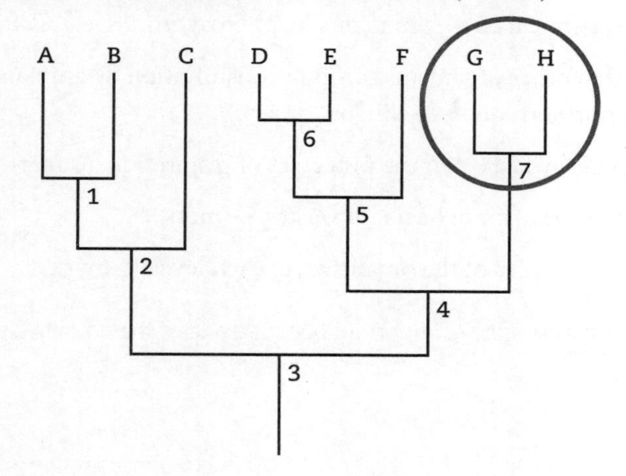

(1 mark)

b Node 4 (1 mark)

c D and E (1 mark)

Question 15

a Symbiosis is an interspecific interaction (1 mark) in which the species live together in a long-term relationship. (1 mark)

b Mutualism means that both species benefit from the interaction. (1 mark)

An example is the relationship between zooxanthellae (photosynthesis producing nutrients for coral) and coral polyps (hosting the zooxanthellae, providing home). (1 mark)

CHAPTER 2: UNIT 3 TOPIC 2

Multiple-choice questions

Question 1 D

A and B start with only one species, therefore no competition. C has both species alive. D is the only option that shows one species surviving at the expense of another.

Question 2 A

Food chains B and C have a tertiary consumer. Food chain D has a quaternary consumer. A is the only one that finishes as a secondary consumer.

Question 3 C

There are no food chains where the northern quoll consumes a plant (TL1); therefore, it cannot occupy TL2 so B and D are incorrect. The northern quoll consumes TL2 organisms and the TL3 organism (the lizard). Therefore, C is correct. A is incorrect because it shows only one trophic level.

Question 4 A

Pandanus is a plant that photosynthesises. None of B, C or D can manufacture their own nutrients.

Question 5 D

Spear grass → insects → small bird → northern quoll → feral cat → dingo (six trophic levels).

Question 6 C

Biomass increases with a climax community, so A and D are incorrect. K-selected species would also increase as the environment becomes more stable, so C is correct and B is incorrect.

Short-response questions

Question 7

Light energy is converted into chemical energy in plants by photosynthesis, e.g. *Pandanus*. (1 mark)

Link to food web made.

Chemical energy is transferred to other organisms through consumption by animals, e.g. *Pandanus* → cockatoo → northern quoll → dingo. (1 mark)

At each level in the food web, energy is used in the processes of respiration. (1 mark)

At each transfer, energy is lost in the form of heat and wastes. (1 mark)

Energy also remains stored in the bodies of the organisms at each level. (1 mark)

The ORDER in which you write this response is important because one of the cognitive verbs is 'sequence'.

Question 8

a 1 mark for substitution:

$$\text{efficiency} = \frac{446}{4633} \times 100 = 9.63\% = 9.6\%$$

1 mark for answer correctly rounded to 1 d.p.

b Any one of the following for 1 mark:

- Energy is used by snails/insect larvae for respiration to survive.
- Energy is lost as waste from snails/insect larvae.
- Not all the snails/insect larvae are consumed by TL3 organisms.

Question 9

10 – 6.3 – 0.4 = 3.3 kg (1 mark)

Working is not required because the cognitive verb is 'determine'.

Question 10

A pyramid of energy shows how much **energy there is in the form of biomass** at a trophic level. The first trophic level always has the largest amount of energy. Therefore, forest and grassland will look the same. (1 mark)

One tree has a much larger biomass than one grass plant. Therefore, the number of trees can be much less than the number of grass plants. (1 mark)

Question 11

a Primary succession. (1 mark) There is no soil present OR substrate is rock OR lichen is only lifeform present. (1 mark)

b (1 mark for each correct pair in a row)

Attribute of succession	Now	In 50 years
r-selected or K-selected species?	r-selected	K-selected
Biodiversity low or high?	Low	High
Biomass low or high?	Low	High

c Now: primary colonisers are small but with higher reproduction rates (r-selected). There are few of them; therefore, fewer organisms to feed on them (low biomass and low biodiversity). (1 mark)

50 years: closer to climax community, larger plants with slower reproduction rates (K-selected and higher biomass). There are more habitats and microhabitats for a greater range of organisms (high diversity). (1 mark)

d Any three of the following for 1 mark each:

- wide tolerance limits
- high seed dispersal rates
- nitrogen-fixers
- high light intensity required
- dormancy.

Question 12

a A keystone species is a plant or animal that plays a **unique and crucial role** in the way an **ecosystem functions**. (1 mark)

b Dingoes prey on kangaroos, feral cats, goats and red fox, reducing their numbers. (1 mark)

Reducing the number of herbivores (kangaroo, goat) means that the grass and herb biomass would increase. (1 mark)

OR

Reducing the number of omnivore (mouse, reptile, bird) predators means more animals survive to feed on grass and herbs, reducing their biomass. (1 mark)

c (1 mark per cycle)

Carbon cycle	Combustion of grasses/herbs returns CO_2 to the atmosphere. OR Carbon is transferred as animals consume plants or other animals.
Water cycle	Water is taken up by plants to be used in photosynthesis or transpiration. OR Water percolates through the soil.

Make sure you refer to evidence presented in the food web.

Question 13

a **i** Bettong: logistic growth (1 mark)

ii Rat: exponential growth (1 mark)

S or J are not correct because these are the curve shapes.

b Density dependent means the effect of the interaction or factor is stronger the greater the density of the population. (1 mark)

c 200 individuals (read from the graph) (1 mark)

d No – the bandicoot population is also increasing as the bettong population increases. (1 mark)

For competitive exclusion, the bandicoot population would be decreasing as the bettong increases. (1 mark)

e Reintroduce a small number of natural predators. (1 mark)

OR

Introduce a second competitor that may be more effective than the stick-nest rat. (1 mark)

Question 14

a A fundamental niche is the specific range of environmental roles, conditions and biological interactions required for survival and reproduction. (1 mark) A realised niche is the actual range of the species, which may be smaller than the fundamental niche. (1 mark)

b No. The graph shows that biological control measures were first implemented around 1955. (1 mark) By this time, the rabbits had spread from Victoria to New South Wales, South Australia, much of Queensland, Western Australia and Tasmania, and half the Northern Territory. (1 mark)

OR

Yes. The graph shows that biological control measures were first implemented around 1955. (1 mark) After 1910/1920 on the map, there has a been relatively small expansion of rabbits into the northern part of Australia. (1 mark)

c The graph shows that each biological control event dramatically reduced the population of rabbits, limiting them to fewer than 20 individuals (or correct value based on graph). (1 mark) Had this reduction in rabbit population not happened, there would probably be more than 75 threatened plant species (from extract) or it may have caused the extinction of these threatened species as well as other species, reducing biodiversity further. (1 mark)

Question 15

Initially, only mosses are present (a type of pioneer species found on rocks). (1 mark) From 15 500 years before present, there are grasses, suggesting soil formation (1 mark), followed by the presence of trees 10 000 years before present, indicating a climax community. (1 mark)

Question 16

Primary succession (e.g. after a volcanic eruption) begins with a bare site that has not been colonised before, whereas secondary succession (e.g. a forest after a fire) occurs in an environment that was previously colonised, but disturbed or damaged. (2 marks)

A second difference is that in primary succession, a pioneer community is required to make the habitat fertile, whereas in secondary succession, the habitat is fertile with soil, seeds and remnants of vegetation. (2 marks)

Question 17

a Producers: $52 - 16 - 8 = 28\,\text{MJ}\,\text{m}^{-2}\,\text{year}^{-1}$ (1 mark)

Herbivores: $28 - 20 - 5 = 3\,\text{MJ}\,\text{m}^{-2}\,\text{year}^{-1}$ (1 mark)

b Respiration was higher for the herbivores. (1 mark)

Decomposition was higher for the producers. (1 mark)

Alternative answer:

- Determines % loss at each trophic level

  $\frac{24}{52} = 46\%$ for producers

  $\frac{25}{28} = 89\%$ for herbivores (1 mark)
- Identifies % loss is higher for herbivores than autotrophs (1 mark)

Question 18

Initially, there is a rapid increase in growth. (1 mark)

This is followed by a sudden drop in population numbers. (1 mark)

This pattern is typical for J-curve population growth. (1 mark)

CHAPTER 3: UNIT 3 DATA TEST

Data set 1

Question 1

Hint
Write down what you know.

$N = ?$

$n = 10$

$M = 26$

$m = 4$

$N = \frac{26 \times 10}{4}$ (1 mark for showing working for the 'calculate' verb)

$N = 65$ (1 mark for correct answer)

Question 2

Before: logistic (1 mark)

After: exponential (1 mark)

Tip (before): graph is almost S-shaped.
Tip (after): graph is J-shaped.

Question 3

Key point: As the population of feral cats has increased/expanded, the population of bandicoots has reduced/constricted. (1 mark)

Question 4

Key point: The highest density areas of feral cats on the map in Figure 3.2 overlap with the historical distribution of the southern brown bandicoot shown in Figure 3.3. (1 mark)

Key point: The current reduced distribution of the southern brown bandicoot in Figure 3.3 could be due to the presence of these predators. (1 mark)

It is important to communicate these key points and refer specifically to the data.

Question 5

Key point: When the fox (predator) was removed, the bandicoot population increased (Figure 3.1). (1 mark)

Key point: Therefore, if the feral cats are removed, the population distribution should increase because there is no longer an overlap with the current or historical distribution of bandicoots. (1 mark)

Data set 2

Question 1

$\Sigma n(n-1) = 11(10) + 12(11) + 4(3) + 4(3) = 266$ (1 mark)

$N(N-1) = 31(30) = 930$ (1 mark)

$\text{SDI} = 1 - \frac{266}{930} = 0.71$ (1 mark)

Question 2

Beach spinifex (1 mark)

Note: There are 19 beach spinifex plants.

Question 3

Pigface, beach spinifex, beach morning glory, beach fan flower. (1 mark for all correct)

OR

Pigface, beach spinifex, beach fan flower, beach morning glory.

OR

Pigface, beach spinifex, beach fan flower and morning glory.

Question 4

Density of pigface at site 1 is $2.4\,\text{m}^{-2}$, which is greater than the density of $1.2\,\text{m}^{-2}$ at site 2. (1 mark)

Refer to data from the tables. State the similarity, difference and significance.

Question 5

Both sites have four different species; therefore, they both have a species richness of 4. (1 mark)

However, they have a different species evenness. Site 2 has predominantly beach spinifex and only a few of the others and could easily lose either beach morning glory or beach fan flower as they only have one because there is only one of each plant. (1 mark)

This means that site 1 has a greater SDI than site 2 and is likely to be more resistant to reduced diversity. (1 mark)

Data set 3

Question 1

Opossums (1 mark)

This group's divergence is from the common ancestor.

Question 2

37–38 million years ago (1 mark)

Look for the most recent common ancestor and look at the time scale for that node.

Question 3

Sugar glider, honey possum, brushtail possum (1 mark)

OR

Honey possum, sugar glider, brushtail possum (1 mark)

Honey possums and sugar gliders have the most recent common ancestor.

Question 4

Pygmy possums and ringtail possums had a common ancestor at 50 mya. Pygmy possums and Tasmanian devils had a common ancestor at approx. 65 mya. (1 mark for both points total)

Therefore, pygmy possums are more closely related to ringtail possums. (1 mark)

Question 5

Opossums (1 mark)

This is because they are the outgroup and first group to branch from the common ancestor. (1 mark)

Note that the question is worth 2 marks; therefore, you should give the reason for your prediction.

Question 6

Since the common ancestor at 50 mya, the ringtail possum clade branches around 28 mya, whereas the pygmy possum clade branches at around 40 mya. (1 mark)

The ringtail possum develops into five species/groups, whereas the pygmy possum develops into two species/groups. (1 mark)

Question 7

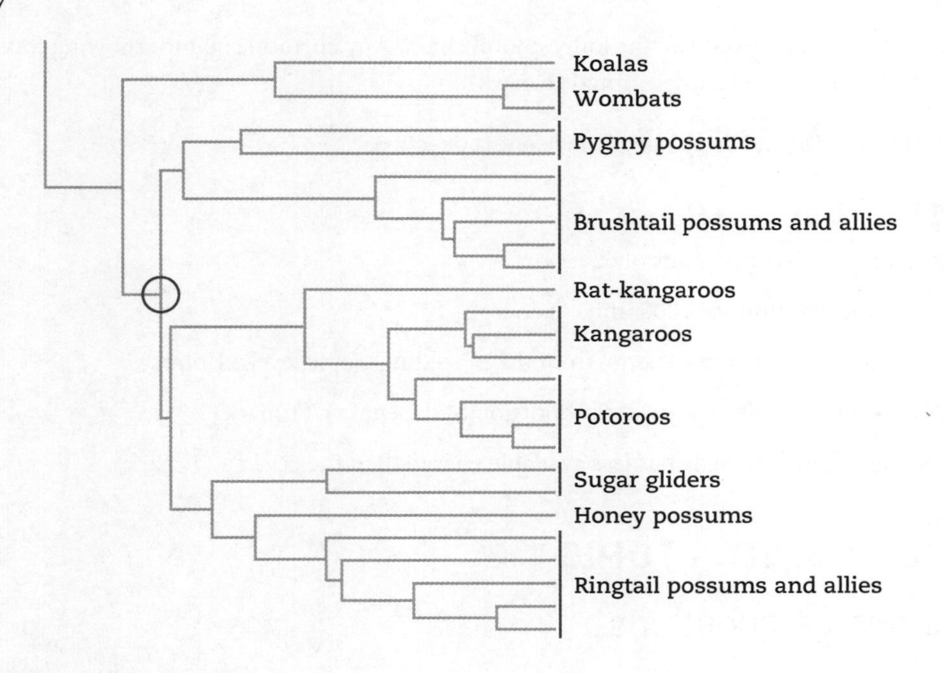

Question 8

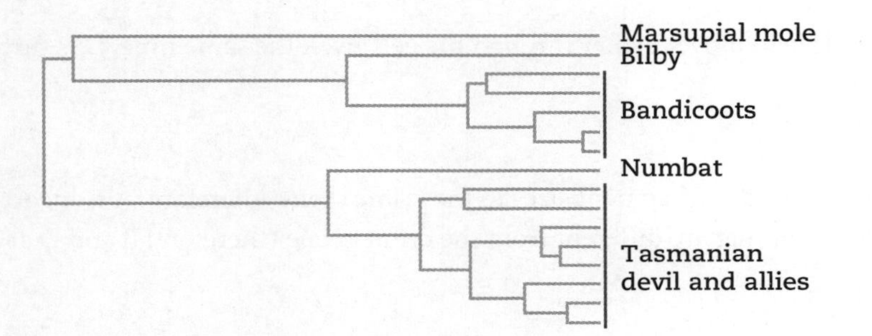

You must include all the lineal descendants from the most recent common ancestor.

Data set 4

Question 1

Primary consumers and decomposers or energy in = 200

Secondary consumers or energy transferred = 30

Efficiency = $\frac{30}{200} \times 100$ (1 mark)

= 15% (1 mark)

Hint
Write what you know.

Question 2

Grass and wattle (1 mark for both)

Question 3

TL2: emu (1 mark)

TL3: echidna, lizard, dingo (1 mark)

TL4: dingo (1 mark)

Question 4

TL1 → TL2 → TL3 → TL4 would be the longest food chain. Any correct sequence showing four trophic levels, e.g. grass → cricket → lizard → dingo (1 mark)

Arrows must be included and be pointing in the correct direction.

Refer to data in the data set. State the similarity, difference and the significance.

Question 5

Eagle has received 3 units of 30 units of energy.

Cricket has received 200 units of 1000 units of energy.

Similarity: Both have received less energy from the preceding tropic level. (1 mark)

Difference: Cricket has received a greater proportion of the energy. (1 mark)

Significance: This means the eagle has less available energy than the cricket. (1 mark)

CHAPTER 4: UNIT 4 TOPIC 1

Multiple-choice questions

Question 1 D

Small fragments are able to move further through the gel, given the same time. D is furthest from where the samples started.

Question 2 C

Suspect 2 has an exact match of fragment sizes to the crime scene (therefore, A is incorrect). Suspects 1 and 3 have fragment sizes that are different from the crime scene (therefore, B and D are incorrect).

Question 3 A

C and D are incorrect because amino acids are protein monomers. B is incorrect because the sugar in DNA or RNA is not sucrose.

Question 4 B

Covalent bonds result from permanently shared electron pairs. Ionic bonds require permanently charged particles called cations and anions. Metallic bonds require a sea of delocalised electrons around metal atoms. These are not present in DNA molecules; therefore, A, C and D are incorrect.

Question 5 D

There are the same number of base pairs in the image; therefore, A and B are not correct. Non-disjunction occurs during meiosis and involves chromosomes; therefore, C is not correct.

Question 6 B

Every human body cell contains the same DNA as 46 chromosomes. The role of cells is determined by how and when genes are expressed; therefore, A, C and D are incorrect.

Question 7 C

DNA needs to be in fragments for the processes of PCR and gel electrophoresis; therefore, C is correct. Genes and genomes are too large. Specific mutation sites are potentially found through these processes.

Question 8 B

Homozygous parents will always produce 100% heterozygous offspring for the dominant trait when the phenotype is controlled by a single gene. Therefore, all the F_1 are heterozygous, e.g. Aa. An F_1 self-cross will produce the genotype ratio AA : 2Aa : aa with the phenotype ratio 3 dominant : 1 recessive. The dominant phenotype is the one expressed in the F_1, which is striped. Therefore, the F_2 will be 3 striped: 1 green.

Short-response questions

Question 9

1 mark for each correct explanation.

N: the mRNA copy/transcript of the DNA leaves the nucleus for the cytoplasm.

O: DNA helix has been unwound and separated; it can be copied or transcribed into mRNA.

P: the amino acids are joined together to form a polypeptide.

Q: the tRNA anticodon binds to the mRNA codon.

R: the mRNA molecule binds to a ribosome.

S: the ribosome identifies the codons in the mRNA sequence.

Hint
Remember: the syllabus statements are very clear about the terms you should use here for each of these statements!

Question 10

a The purpose of gel electrophoresis is to separate molecules of DNA, RNA or proteins (1 mark) on the basis of size, shape and charge (1 mark).

b 6.5 kb (1 mark)

Look at the gel. See how long each fragment in lane E is, then add the numbers together.

c 3 sites (1 mark)

Remember: plasmids are circular DNA; therefore, 3 pieces = 3 cuts.

d Lane D. (1 mark) There are two sites, so two fragments. (1 mark) Fragments are 10 kb and 6 kb in length. (1 mark)

e To make recombinant DNA, the target DNA is isolated from a cell. (1 mark) It is cut by a restriction enzyme. The same restriction enzyme is used to cut a plasmid. (1 mark) The target DNA is inserted into the plasmid. (1 mark) The transformed plasmid is taken up by bacteria and the target DNA is replicated. (1 mark)

f HindII is the best choice. (1 mark) It has the same recognition sequence for cutting as the target gene. (1 mark) Therefore, the target gene and plasmid will have the same sticky ends for joining. (1 mark)

Question 11

a Either cell that looks like these: (1 mark)

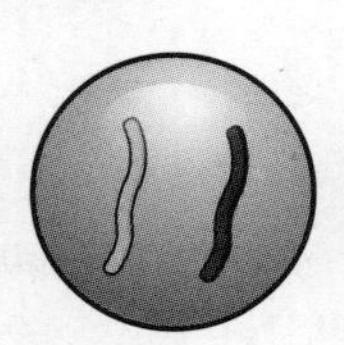
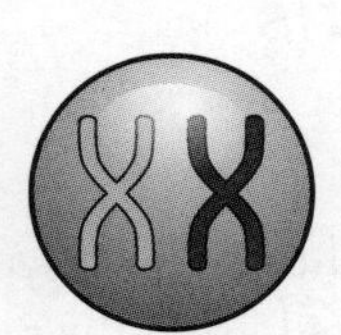

b 1 mark for each similarity/difference. 1 mark for each relevant significance statement. Must be based on the diagram.

Similarity	Significance	Difference	Significance
Gametes are haploid or have '*n*' chromosomes	Results in a diploid zygote after fertilisation	4 sperm but only 1 egg	Sperm have to move to the egg. Need higher numbers to ensure some survive. Egg already at fertilisation site
Gametes are not identical	Introduced variation into offspring	Egg is large, sperm is small	Egg provides nutritional support after fertilisation. Sperm provides only nucleus
		Sperm have tail, egg does not	Sperm need to move independently to reach egg. Egg not required to move to reach sperm

Question 12

a X-linked. (1 mark) All the males have only one band (XY) on the gel, whereas the females have 2 bands (XX). (1 mark)

b Four (there are four shaded shapes). (1 mark)

c No, he will not have the disorder. (1 mark). Has only one band at 26 repeats. This is less than the 36 repeats required for the disorder to be present. (1 mark)

Hint
Think: the mother must be heterozygous to have unaffected sons.

d

	X^H	X^h
X^h	X^HX^h	X^hX^h
Y	X^HY	X^hY

Correct parents (2 marks), correct completion of table. (1 mark)

Phenotypes possible: 1 female with disorder: 1 male with disorder: 1 normal female: 1 normal male. (1 mark)

Question 13

Helicase unzips the DNA molecule by breaking the weak hydrogen bonds between the two complementary strands. (1 mark)

This creates a replication fork region so that bases are exposed. (1 mark)

DNA polymerase uses each original strand as a template to produce a copy of the DNA molecule, and adds complementary nucleotides to the exposed bases. (1 mark)

DNA polymerase also proofreads the newly synthesised strand. (1 mark)

Question 14

Transcription involves copying a gene's DNA sequence to make an RNA molecule. (1 mark) This is performed by enzymes called RNA polymerases, which link nucleotides to form an mRNA strand. (1 mark)

In the translation process, the mRNA formed in transcription is transported out of the nucleus to the ribosome. (1 mark) Here, it directs protein synthesis. The mRNA passes through the ribosome and tRNA interacts with it, adding amino acids together to make a protein chain. (1 mark)

CHAPTER 5: UNIT 4 TOPIC 2

Multiple-choice questions

Question 1 B

The mean has moved to the right of the original mean, shifting the curve in one direction: therefore, B is correct. Disruptive selection (C) would split the phenotype into two extremes. Stabilising selection (D) would have the original mean at a higher frequency. Destabilising selection (A) is not an accepted description associated with phenotypic selection.

Question 2 C

The original population (grey) has lost the middle range of antennae length and only the extremes are presented (black). Therefore, A, B and D are incorrect. See explanation for Question 1.

Question 3 D

Rat and mouse have the highest number of similarities with no differences.

Question 4 B

Of the pairs provided, the frog and the pig have the greatest number of differences with 5. The dog and pig have 1 (A); the pig and rat have 3 (C), and the rat and mouse have none (D).

Question 5 A

A is correct because the Ordovician has the greatest increase in marine family numbers (+300) – from approximately 150 to 450. C and D don't have observable increases. The Cretaceous (B) has a smaller increase from approximately 400 to 610 (+210).

Question 6 D

Microevolution is small-scale variation. When a group accumulates enough variation due to microevolution, it may become a macroevolutionary change. As microevolution results from mutation, genetic drift and gene flow, macroevolution depends on the same things.

Short-response questions

Question 7 (1 mark)

Change in genetic composition of a population during successive generations, which may result in the development of a new species.

Key points: change in genes, many generations, new species

Question 8

Organisms W and Y are most closely related because the DNA sequence for this gene is the same. (1 mark)

Organism X is the next most closely related organism. There is one change from W and Y at the eighth base from A to C. (1 mark)

Organism Z is the most distantly related because there are six changes from organisms W and Y. (1 mark)

Question 9

a Predation by the frog. (1 mark)

b Negative because the a allele is reducing in frequency. (1 mark)

Question 10

a Founder effect. (1 mark) A small number of individuals separated to form a new population rather than a dramatic reduction in population size. (1 mark)

b The transplanted ladybugs have a non-random selection of alleles (fewer A alleles) (1 mark) and they also have a mutant allele (A*) that is not present in the original population (1 mark), changing the allele frequency from the original population.

c No, because of temporal isolation. (1 mark) If the ladybugs are not present at the same time in the environment, they cannot breed, so there would be no gene flow. (1 mark)

Question 11

a Allopatric speciation. (1 mark) Gene flow into Sukla is small and they are geographically separated from the other populations. (1 mark)

Not parapatric because there is no overlap in the populations.

b The combination of small population size (smallest circle) (1 mark) and low gene flow (narrow arrow) (1 mark) and genetic drift (1 mark) has a greater effect in reducing genetic diversity.

Therefore, they may not be able to respond to selection pressures. (1 mark)

Remember to say that the genetic diversity is much less!

c Clearing of forests OR building of roads. (1 mark)

d Chitwan is 450 km from Sukla, whereas there is only 136 km between Sukla and Bardia, and 314 km between the larger populations of Bardia and Chitwan. (1 mark)

Question 12 (2 marks)

Scenario 1: coevolution (1 mark)

Scenario 2: divergent evolution (1 mark)

Scenario 1: one organism is dependent on another for reproduction

Scenario 2: a trait persisting in different forms

Question 13

In a population, some individuals will have inherited traits that help them survive and reproduce. (1 mark) Because the helpful traits are heritable, and because organisms with these traits leave more offspring, the population will become adapted to its environment. (1 mark)

In the case of the lungfish, if the environment remains relatively unchanged (i.e. no new predators or competitors, still a water-dwelling organism), there is no environmental selection pressure to select for any new mutations in morphology, so there is minimal change in the species. (1 mark)

Question 14

a Directional selection is a mode of natural selection in which an extreme phenotype is favoured over other phenotypes (1 mark), causing the allele frequency to shift over time in the direction of that phenotype. (1 mark)

b

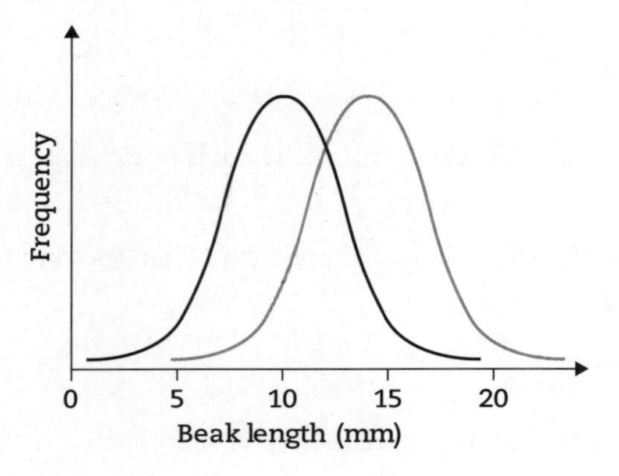

Question 15

At time point I, there is equal gene flow and equal allelic frequency in all niches, indicating a high degree of interbreeding between all niches.

Trends show that the niche labelled as D has a progressive decrease in allelic frequency of the gene from time point I to time point III. This is supported by gene flow halting between C and D at time point II and then further from B to D at time point III.

Gene flow between niches A, B and C remains constant throughout all time points, as shown by the arrows and by the constant allelic frequency.

This evidence supports a potential speciation event at niche D.

However, niche D is not totally isolated because there remains some gene flow to the other populations (through niche A). This excludes allopatric speciation as the mode of proposed speciation.

However, there is an element of population isolation through niches, which supports parapatric speciation over sympatric speciation.

For 6 marks, the response:

- identifies three pieces of evidence of speciation
- infers that speciation is not allopatric because D is not isolated, e.g. at III, there is still gene flow with niche A
- concludes that parapatric speciation occurred at D.